THE

REASON WHY:

GENERAL SCIENCE.

A CAREFUL

COLLECTION OF MANY HUNDREDS OF REASONS FOR THINGS WHICH, THOUGH GENERALLY BELIEVED, ARE IMPERFECTLY UNDERSTOOD.

A BOOK OF

Condensed Scientific Knowledge for the Million.

BY THE AUTHOR OF

"INQUIRE WITHIN," "THE BIBLICAL REASON WHY," "THE REASON WHY: NATURAL HISTORY," "THAT'S IT, OR PLAIN TEACHING," "THE CORNER CUPBOARD."

This collection of useful information on "Common Things" is put in the interesting form of "Why and Because," and comprehends a familiar explanation of many subjects which occupy a large space in the philosophy of Nature, relating to air, animals, atmosphere, caloric, chemistry, ventilation, materia medica, meteorology, acoustics, electricity, light, zoölogy, etc.

NEW YORK:

DICK & FITZGERALD, PUBLISHERS

No. 18 ANN STREET.

PREFACE.

We are all children of one Father, whose Works it should be our delight to study. As the intelligent child, standing by his parent's knee, asks explanations alike of the most simple phenomena, and of the most profound problems; so should man, turning to his Creator, continually ask for knowledge. Not because the profession of letters has, in these days, become a fashion, and that the man of general proficiency can best work out his success in worldly pursuits; but because knowledge is a treasure which gladdens the heart, dignifies the mind, and ennobles the soul.

The occupation of the mind, by the pursuit of knowledge, is of itself a good, since it diverts from evil, and by elevating and refining the mind, and strengthening the judgment, it fortifies us for the hour of temptation, and surrounds us with barriers which the powers of sin cannot successfully assail.

It is not contended that the mere acquisition of knowledge will either ensure a good moral nature, or convey religious truth. But both religion and morals will find in the diffusion of knowledge a ground-work upon which their loftier temples may discover an acceptable foundation.

The man who comprehends the order of Nature, and the immutability of Divine law, must of necessity bring himself in some degree into accordance with that order, and under submission to the law: hence the *tendency* of knowledge will always be found to harmonise the fragment with the mass, and to subvert the evil to the good.

The troubles of the world have arisen from the want of knowledge, not from the possession of it. And in proportion as man becomes an intelligent and reflective being, he will be a better creature in all the relations of life. If these benefits, vast and incalculable as they are, be the real tendency and result of knowledge, why is ignorance allowed to remain, and why is the world still distracted by error?

It is because the moral and intellectual qualities of man are, like all creations and gifts of God, the subjects of development, whose law is progression.

We can aid human improvement, but we cannot unduly hasten it. Whenever man has sprung too rapidly to a conclusion, he has alighted upon error, and has had to retrace his steps.

The greatest philosophers have been those who have clung to the demonstrative sciences, and have held that a simple truth well ascertained, is greater than the grandest theory founded upon questionable premises. Newton made more scientific revelations to mankind than any other philosopher; and his discoveries have borne the searching test of time, because he snatched at nothing, leaped over no chasm to establish a favourite dogma; but, by the slowest steps, and by regarding the merest trifles, as well as the highest phenomena, he learnt to read Nature correctly. He discovered that her atoms were letters, her blades of grass were words, her phenomena were sentences, and her complete volume a grand poem, teaching on every page the wisdom and the power of an Almighty Creator.

When he observed an apple fall to the ground, he asked the "Reason Why;" and in answer to that enquiry, there came one of the grandest discoveries that has ever been recorded upon the book of science. With that discovery a flood of light burst upon the human mind, illustrating in a far higher degree than had ever previously been conceived, the vastness of Almighty Power.

Why should not each of us enquire the "Reason Why" regarding everything that we observe? Why should we montally *grope* about, when we may *see* our way? When addressed in a foreign tongue, we hear a number of articulated sounds, to which we can attach no meaning; they convey nothing to the mind, make no impression upon the indwelling soul. When those sounds are interpreted to us, in a language that we can understand, they impart impressions of joy, hope, surprise, or sorrow, because the words convey to us *a meaning.* In like manner, if we fail to understand Nature, its beauties, its teachings are lost. Everything speaks to us, but we do not understand the voices. They come murmuring from the brook, trilling from the bird, or pealing from the thunder; but though they reach the ear of the body, they do not impress the listening spirit.

Every flower, every ray of light, every drop of dew, each flake of snow, the curling smoke, the lowering cloud, the bright sun, the pale moon, the twinkling stars, speak to us in eloquent language of the great Hand that made them. But millions lose the grand lesson which Nature teaches, because they can attach no meaning to what they see or hear.

"THE REASON WHY" is offered as an interpreter of many of Nature's utterances. Great care has been taken that these interpretations may be consistent with the latest knowledge, obtained from the highest sources. If the author finds that his work is accepted for the good of those who seek not only to know, but to *understand,* he will make it his constant care to read the Book of Nature, and to add to the pages of this volume whatever interpretations the progress of enquiry and discovery may demand and supply.

INDEX, AND INDEX LESSONS.

☞ *The numbers refer to the Questions. The Index Lessons do not correspond with the Chapters, but are designed to bring together in their alphabetical connection, all the Questions and Answers upon each particular subject included in the work.*

LESSON I.

LESSON II.

LESSON III.

LESSON IV.

LESSON V.

LESSON VI.

LESSON VII.

LESSON VIII.

LESSON IX.

LESSON X.

LESSON XII.

LESSON XI.

LESSON XIII.

LESSON XIV.

LESSON XV.

LESSON XVI.

LESSON XVII.

LESSON XVIII.

LESSON XIX.

LESSON XX.

LESSON XXI.

LESSON XXII.

LESSON XXIII.

LESSON XXIV.

LESSON XXV.

LESSON XXVI.

LESSON XXVII.

LESSON XXVIII.

LESSON XXIX.

LESSON XXX.

LESSON XXXI.

LESSON XXXII.

LESSON XXXIII.

LESSON XXXIV.

LESSON XXXV.

LESSON XXXVI.

LESSON XXXVII.

LESSON XXXVIII.

LESSON XXXIX.

LESSON XL.

LESSON XLI.

LESSON XLII.

LESSON XLIII.

LESSON XLIV

LESSON XLV.

LESSON XLVI.

LESSON XLVII.

LESSON XLVIII.

LESSON XLIX.

LESSON L.

LESSON LI.

LESSON LII.

LESSON LIII.

LESSON LIV.

LESSON LV.

LESSON LVI.

LESSON LVII.

LESSON LVIII.

LESSON LIX.

LESSON LX.

LESSON LXI.

LESSON LXII.

LESSON LXIII.

LESSON LXIV.

LESSON LXV.

LESSON LXVI.

LESSON LXVII.

LESSON LXVIII.

LESSON LXIX.

LESSON LXX.

LESSON LXXI.

LESSON LXXII.

LESSON LXXIII.

LESSON LXXIV.

LESSON LXXV.

LESSON LXXVI.

LESSON LXXVII.

LESSON LXXVIII.

"God looked down from heaven upon the children of men, to see if there were any that did understand, that did seek God."—PSALM LIII.

THE REASON WHY:

GENERAL SCIENCE.

CHAPTER I

1. *Why should we seek knowledge?*

Because it assists us to comprehend the *goodness and power of God.*

And it gives us power over the circumstances and associations by which we are surrounded: the proper exercise of this power will greatly promote our happiness.

2. *Why does the possession of knowledge enable us to exercise power over surrounding circumstances?*

Knowledge enables us to understand that, in order to live healthily, we require to breathe fresh and pure air. It also tells us that animal and vegetable substances, undergoing decay, poison the air, though we may not be able to see, or to smell, or otherwise discover the existence of such poison. Knowing this, we become careful to remove from our presence all such matters as would tend to corrupt the atmosphere. This is only one of the countless instances in which knowledge gives us power over surrounding circumstances.

3. *Name some other instances in which knowledge gives us power.*

Knowledge of *Geography* and of *Navigation* enables the mariner to guide his ship across the trackless deep, and to reach the sought-for port, though he had never before been on its shores.

Knowledge of *Chemistry* enables us to separate or to combine the various substances found in nature. Thus we obtain useful and precious metals from what at first appeared to be useless stones transparent glass from pebbles, through which no light could pass; soap from oily substances; and gas from solid bodies.

Knowledge of *Medicine* enables the physician to overcome the ravages of disease, and to save suffering patients from sinking prematurely to the grave.

Knowledge of *Anatomy* and of *Surgery* enables the surgeon to bind up dangerous fractures and wounds, and to remove, even from the internal parts of bodies, ulcers and diseased formations that would otherwise be fatal to life.

Knowledge of *Mechanics* enables man to increase his power by the construction of machines. The steam-ship crossing the ocean in opposition to wind and tide, the railway locomotive travelling at 60 miles an hour, and the steam-hammer beating blocks of iron into useful shapes, are evidences of the power which man acquires through a knowledge of mechanics.

Knowledge of *Electricity* enables man to stand in comparative safety amid the awful war of the elements. Lightning, the offspring of electricity, has a tendency to strike upon lofty objects by which it may be attracted. By its mighty powers churches or houses may be instantly levelled with the dust. But man, knowing that electricity is strongly attracted by particular substances, raises over lofty buildings rods of steel communicating with bars that descend into the ground. The lightning, rushing with indescribable force toward the steeple, is attracted by the bar of steel, and conducted harmlessly to the earth. Man may thus be said to take even lightning by the hand, and to divert its destroying force by the aid of Knowledge. And in countless other instances "Knowledge is Power."

CHAPTER II.

4. *Why do we breathe air?*

Because the air contains *oxygen*, which is necessary to life.

5. *Why is oxygen necessary to life?*

Because it combines with the *carbon* of the blood, and forms *carbonic acid gas.*

"Be not as the horse, or as the mule, which have no understanding: whose mouth must be held with the bit and bridle."—PSALM XXXII.

6. *Why is this combination necessary?*

Because we are so created that the substances of our bodies are constantly undergoing change, and this resolving of solid matter into a gaseous form, is the plan appointed by our Creator to remove the matter called *carbon* from our systems.

7. *Why do our bodies feel warm?*

Because, in the union of *oxygen* and *carbon*, heat is developed.

8. *What is this union of oxygen and carbon called?*

It is called *combustion*, which, in chemistry, means the decomposition of substances, and the formation of new combinations, accompanied by heat; and sometimes by light, as well as heat.

9. *What is formed by the union of oxygen and carbon?*

Carbonic acid gas.

10. *What becomes of this carbonic acid gas?*

It is sent out of our bodies by the compressure of the lungs, and mingles with the air that surrounds us.

11. *Is this carbonic acid gas heavier or lighter than the air?*

Pure carbonic acid gas is the heaviest of all the gases. That which is sent out of the lungs is not pure, because the whole of the air taken into the lungs at the previous inspiration has not been deprived of its *oxygen*, and the nitrogen is returned. Therefore the breath sent out of the lungs may be said to consist of *air*, with a large proportion of *carbonic acid gas*.

12. *What is the composition of air in its natural state?*

It consists of *oxygen*, *nitrogen*, and *carbonic acid gas*, in the proportions of oxygen 20 volumes, nitrogen 79 volumes, and carbonic acid gas 1 volume. It also contains a slight trace of watery vapour.

13. *What is the state of the air after it has once been breathed?*

It has parted with about one-sixth of its oxygen, and taken up an equivalent of carbonic acid. And were the same air to be breathed six times successively, it would have parted with *all* its oxygen, and could no longer sustain life.

14. *Is the impure air sent out of the lungs lighter or heavier than common air?*

At first, being rarefied by warmth, it is *lighter*. But, if undisturbed, it would become *heavier* as it cooled, and would descend.

15. *Why is it proper to have beds raised about two feet from the ground?*

Because at night, the bed-room being closed, the breath of the sleeper impregnates the air of the room with carbonic acid gas, which, descending, lies in its greatest density near to the floor.

16. *What are the chief sources of carbonic acid gas?*

The vegetable kingdom (as will be hereafter explained), the combustion of substances composed chiefly of carbon, the breathing of animals, and the decomposition of carbonic compounds.

17. *Is breathing a kind of combustion?*

It is. In the breathing of animals, the burning of coals, or of wood, or candles, &c., similar changes occur. The *oxygen* of the air combines with the *carbon* of the substance said to be burnt, and forms *carbonic acid gas*, which unfits the air for the purposes of either breathing or of burning, until it has been renewed by admixture with the air.

18. *What is carbon?*

It is one of the elementary bodies, and is very abundant throughout nature. It abounds mostly in vegetable substances, but is also contained in animal bodies, and in minerals. The form in which it is most familiar to us is that of *charcoal*, which is carbon almost pure.

19. *What is meant by an elementary body?*

An elementary body is one of those substances in which chemistry is unable to discover more than one constituent. For instance, the chemist finds that water is composed of *oxygen* and *hydrogen*. Water is therefore a *compound* body. But *carbon* consists of *carbon only*, and therefore it is called a simple, or elementary body.

20. *Why is it dangerous to burn charcoal in rooms?*

Because, being composed of *carbon* that is nearly pure, its combustion gives off a large amount of *carbonic acid gas.*

21 *What is the effect of carbonic acid gas upon the human system?*

It induces drowsiness and stupor, which, if not relieved by ventilation, would speedily cause death.

22. *What is the reason that people feel drowsy in crowded rooms?*

Because the large amount of carbonic acid gas given off with the breaths of the people, makes the air poisonous and oppressive.

23. *What other causes of drowsiness are there?*

The candles, gas, or fires that may be burning in the rooms where people are assembled. Three candles produce as much carbonic acid gas as one human being; and it is probable that one gas-light produces as much carbonic acid gas as two persons.

24. *Have people ever been poisoned by their own breaths?*

In the reign of George the Second, the Rajah of Bengal took some English prisoners in Calcutta, and put 146 of them into a place which was called the "Black Hole." This place was only 18 feet square by 16 feet high, and ventilation was provided for only by two small grated windows. *One hundred and twenty-three of the prisoners died in the night,* and most of the survivors were afterwards carried off by putrid fevers. Many other instances have occurred, but this one is the most remarkable.

CHAPTER III.

25. *What is oxygen?*

Oxygen is one of the most widely diffused of the elementary substances. It is a gaseous body.

26. *Why do persons who are walking, or riding upon horseback feel warmer than when they are sitting still?*

Because as they breathe more rapidly, the combustion of the *carbon* in the blood is increased by the *oxygen* inhaled, and greater heat is developed.

27. *Why does the fire burn more brightly when blown by a bellows?*

Because it receives, with every current of air, a fresh supply of *oxygen*, which unites with the *carbon* and *hydrogen* of the coals, causing more rapid combustion and increased heat.

28. *Why does not the oxygen of the air sometimes take fire?*

Because oxygen, *by itself*, is incombustible. The wick of a candle, which retains the slightest spark, being immersed in oxygen, will instantly burst into a brilliant flame; and even a piece of iron wire made red-hot, and dipped in oxygen, will burn rapidly and brilliantly. Oxygen, though non-combustible of itself, is the most powerful *supporter of combustion*.

29. *Why do we know that oxygen will not burn of itself?*

Because when we immerse a burning substance into a jar of oxygen, it immediately burns with intense brilliancy; but directly it is withdrawn from the oxygen, the intensity of the flame diminishes, and the oxygen which remains is *unaffected*.

30. *Why do we know that oxygen is necessary to our existence?*

Because animals placed in any kind of gas, or in any combination of gases, where oxygen *does not exist*, die in a very short time.

31. *Where is oxygen found?*

It is found in the air, mixed with *nitrogen*; in water combined with *hydrogen*; in the tissues of vegetables and animals; in our blood; and in various compounds called, from the presence of oxygen, *oxides*.

32. *Why is the oxygen of the air mixed so largely with nitrogen?*

Because *oxygen* in any greater proportion than that in which it is found in the atmosphere, would be too exciting to the animal system. Animals placed in *pure oxygen* die in great agony from fever and excitement, amounting to madness.

33. *What is nitrogen?*

Nitrogen is an elementary body in the form of gas.

34. *Where is nitrogen found?*

It is chiefly found in the air, of which it constitutes 79 out of 100 volumes. It may be mixed with oxygen in various proportions; but in the atmosphere it is uniformly diffused. It is found in most animal matter, *except fat and bone.* It is not a constituent of the *vegetable acids*, but it is found in most of the *vegetable alkalies.*

35. *What are acids?*

Acids are a numerous class of chemical bodies. They are generally sour. Usually (though there are exceptions) they have a great affinity for water, and are easily soluble therein; they unite readily with most *alkalies*, and with the various *oxides.* All acids are compounds of two or more substances. Acids are found in all the kingdoms of nature.

36. *What are alkalies?*

Alkalies are a numerous class of substances that have a great affinity for, and readily combine with, *acids*, forming *salts.* They exercise peculiar influence upon vegetable colours, turning blues green, and yellows reddish brown. But they will restore the colours of vegetable blues which have been reddened by acids; and, on the other hand, the *acids* restore vegetable colours that have been altered by the *alkalies.* Alkalies are found in all the kingdoms of nature.

37. *Could animals live in nitrogen?*

No; they would immediately die. But a mixture of *oxygen* and *nitrogen*, in equal volumes, constitutes *nitrous oxide*, which gives a pleasurable excitement to those who inhale it, causing them to be merry, almost to insanity; it has, therefore, been called *laughing gas.*

38. *Why does nitrous oxide produce this effect?*

Because it introduces into the body more *oxygen* than can be consumed. It, therefore, deranges the nervous system, and being a powerful stimulant, gives an unnatural activity to the nervous centres and the brain.

39. *In what proportions are the atmospheric gases found in the blood?*

The mean quantity of the gases contained in the human blood has been found to be equal to 1-10th of its whole volume. In *venous* blood, the average quantity of *carbonic acid* is about 1-18th, that of *oxygen* about 1-85th, and that of *nitrogen* about 1-100th of the volume of the blood. In *arterial* blood their quantities have been found to be *carbonic acid* about 1-14th, *oxygen* about 1-38th, and *nitrogen* about 1-72nd.

40. *Then is nitrogen taken into the blood from the air?*

Such a supposition is highly improbable. It is probably derived from *nitrogenised food,* just as *carbonic acid* is derived from *carbonised food.*

41. *What is venous blood?*

Venous blood is that which is returning through the *veins* of the body from the organs to which it has been circulated.

42. *What is arterial blood?*

Arterial blood is that which is flowing from the heart through the *arteries* to nourish the parts where those arteries are distributed.

43. *What is the difference between venous and arterial blood?*

Venous blood contains *more* carbonic acid, and *less* oxygen and nitrogen than arterial blood.

44. *Will nitrogen burn?*

It will not burn, nor will it support combustion.

45. *What is the difference between "burning" and "supporting combustion?"*

Oxygen gas will not burn of itself, but it aids the decomposition by fire of bodies that are combustible. It is therefore called a *supporter of combustion.* But hydrogen gas, *though it burns of itself* will extinguish a flame immersed in it. It is therefore said to be a body which will *burn*, but *will not support combustion.*

46. *What becomes of the nitrogen that is inhaled with the air?*

It is thrown off with the breath, mixed with *carbonic acid gas*, and flies away to be renewed by a fresh supply of oxygen.

47. *Where does nitrogen find a fresh supply of oxygen?*

In the atmosphere. Nitrogen is said to possess a remarkable tendency to *mix* with oxygen, without having a positive chemical affinity for it. That is to say, neither the *oxygen* nor the *nitrogen* undergoes any change by the union, except that of *admixture.* The oxygen and the nitrogen still possess their own peculiar properties. Oxygen and nitrogen are found in nearly the same proportions in all climates, and at all altitudes.

48. *In combustion does any other result take place besides the union of oxygen and carbon forming carbonic acid gas?*

Yes. Usually *hydrogen* is present, which in burning unites with *oxygen*, and forms *water.*

CHAPTER IV.

49. *What is hydrogen?*

Hydrogen is an elementary gas, and is the lightest of all known bodies.

50. *Will hydrogen support animal life?*

It will not. It proves speedily fatal to animals.

51. *Will hydrogen support combustion?*

Although it will burn, yielding a feeble bluish light, it will, if pure, extinguish a flame that may be immersed in it. Hydrogen will therefore *burn*, but will not *support combustion.*

52. *Why will hydrogen explode, if it will not support combustion?*

When hydrogen explodes it is always in combination with *oxygen*,

"As smoke is driven away, so drive them away: as wax melteth before the fire, so let the wicked perish at the presence of God."—PSALM XLVI.

or with the common air, which contains *oxygen*. *Two* measures of hydrogen and *one* of oxygen form a most explosive compound.

53. *Why does hydrogen explode, when mixed with oxygen, upon being brought in contact with fire?*

Because of its strong affinity for *oxygen*, with which, upon the application of heat, it unites to form water.

54. *Where does hydrogen chiefly exist?*

In the form of *water*, where it exists in combination with *oxygen*. *Eleven* parts of hydrogen, and *eighty-nine* of oxygen, form water.

55. *Is hydrogen found elsewhere?*

It is never found but in a state of combination; united with oxygen, it exists in *water;* with nitrogen, in *ammonia;* with chlorine, in *hydro-chloric acid;* with fluorine, in *hydro-fluoric acid;* and in numerous other combinations.

56. *Is the gas used to illuminate our streets, hydrogen gas?*

It is; but it is combined with carbon, derived from the coals from which it is made. It is therefore called *carburetted hydrogen*, which means *hydrogen* with *carbon*.

57. *How is hydrogen gas obtained from coals?*

It is driven out of the coals by heat, in closed vessels, which prevent its union with *oxygen*.

58. *What becomes of the water which is formed by the burning of hydrogen in oxygen?*

It passes into the air in the form of watery vapour. Frequently it condenses, and may be seen upon the walls and windows of rooms where many lights or fires are burning. Sometimes, also, portions of it become condensed in the globes of the glasses that are suspended over the jets of gas. *A large volume of these gases forms only a very small volume of water.*

59. *What becomes of the carbonic acid gas which is produced by combustion?*

It is diffused in the air which should be removed by adequate ventilation

"I will both lay me down in peace and sleep: for thou, Lord, only, makest me dwell in safety."—PSALM IV.

60. *What proportion of carbonic acid gas is dangerous to life?*

Any proportion over the natural one of 1 per cent. may be regarded as *injurious*. But toxicologists state that *five per cent.* of carbonic acid gas in the atmosphere is *dangerous* to life.

61. *What are toxicologists?*

Persons who study the nature and effects of poisons and their antidotes.

62. *Which kind of combustible used for lighting tends most to vitiate the air?*

Assuming all the lights to be of the same intensity, the degree in which the substances burnt would vitiate the atmosphere may be gathered from the number of minutes each would take to exhaust a given quantity of air. This has been found to be: rape oil, 71 minutes; olive oil, 72; Russian tallow, 75; town tallow, 76; sperm oil, 76; stearic acid, 77; wax candles, 79; spermaceti candles, 83; common coal gas, 98; canal coal gas, 152. Thus it is shown that rape oil is *most destructive* of the atmosphere, and that coal gas is the *least destructive*.

63. *Is an escape of hydrogen gas from a gas-pipe dangerous to life?*

It is dangerous, first, by *inhalation*. There are no less than six deaths upon record of persons who were killed by sleeping in rooms near to which there was a leakage of gas.

It is dangerous, secondly, by *explosion*.

[In 1848, an explosion of gas occurred in Albany-street, Regent's-park, London. The gas accumulated in a shop for a very short time only. It had been escaping from a crack in the meter for about one hour and twenty minutes. The area of the room was about 1,620 cubic feet. When the gas exploded, it blew out the entire front of the premises, carried two persons through a window into an adjoining yard, and forced another person on to the pavement on the opposite side of the street, where she was killed. The effect of the explosion was felt for more than a quarter of a mile on each side of the house, and most of the windows in the neighbourhood were shattered. The iron railings over the area of the house directly opposite were snapped asunder; and a part of the roof, and the back windows of another house, were carried to a distance of from 200 to 300 yards. The pavement was torn up for a considerable length, and the damage done to 103 houses was afterwards reported to amount to £20,000. Other serious explosions have taken place. The explosions of "*coal damp*," which frequently occur in mines, are of a similar character.]

64. *What proportion of hydrogen gas with atmospheric air will explode?*

According to the researches of Sir Humphrey Davy, *seven* or *eight* parts of *air*, to *one* of *gas*, produce the greatest explosive effect; while *larger* proportions of gas are less dangerous. A mixture of *equal parts* of gas and air will burn, but it will not explode. The same is the case with a mixture of *two* of *air*, or *three* of *air*, and *one* of *gas*; but *four* of *air* and *one* of *gas* begin to be explosive, and the explosive tendency increases up to *seven* or *eight* of *air* and *one* of *gas*, after which the increased proportion of gas diminishes the force of the explosion.

65. *What is the best method of preventing the explosion of gas?*

Observe the rule, *never to approach a supposed leakage with a light.* Fortunately the gas, which threatens our lives, warns us of the danger by its pungent smell. The first thing to be done is to open windows and doors, and to ventilate the apartment. Then turn the gas off at the main, and wait a short time until the accumulated gas has been dipersed.

66. *Does hydrogen gas rise or fall when it escapes?*

Being *twelve times lighter than common air* it *rises*, and therefore it would be better for ventilation to open the window at the *top* than at the *bottom*. But all gases exhibit a strong tendency to *diffuse themselves*, and therefore they do not rise or fall in the degree that might be anticipated.

67. *What proportion of hydrogen in the air is dangerous to life, if inhaled?*

One-fiftieth part has been found to have a *serious effect* upon animals. The effects it produces upon the human system are those of depression, headache, sickness, and general prostration of the vital powers. It is therefore advisable to observe precautions in the use of gas.

68. *What proportion of gas in the air may be recognised by the smell?*

By persons of acute powers of smelling it may be recognised when there is *one* part of *gas* in *five hundred parts of atmospheric air;* but it becomes very perceptible when it forms *one* part in *a hundred and fifty.* Warning is, therefore, given to us long before the point of danger arrives.

69. *What other sources of hydrogen are there in our dwellings?*

It arises from the decomposition of animal and vegetable substances, containing *sulphur* and *hydrogen.* These give off a gas called *sulphuretted hydrogen,* from which the fætid effluviam of drains and water-closets chiefly arise. We should, therefore, take every precaution to secure effective drainage, and to keep drain-traps in proper order.

70. *May the use of gas for purposes of illumination be considered highly dangerous?*

Not if it is intelligently managed. The appliances for the regulation of gas are so very simple and perfect, that accidents seldom arise except from neglect. In England 6,000,000 tons of coal are usually consumed in the manufacture of gas, producing 60,000,000,000 cubic feet of gas. And yet accidents are of very uncommon occurrence.

CHAPTER V.

71. *What is heat?*

Heat is a principle in nature which, like light and electricity, is best understood by its *effects.* We popularly call that heat, which raises the temperature of bodies submitted to its influence.

72. *What is caloric?*

Caloric is another term for heat. It is advisable, however, to use the term *caloric* when speaking of the *cause* of heat, and of *heat* as the *effect* of the presence of *caloric.*

"While the earth remaineth, seed-time and harvest, and cold and heat, and summer and winter, and day and night, shall not cease."—GEN. VIII.

73. *What is the source of caloric?*

The sun is its chief source. But caloric, in some degree, exists *in every known substance.*

74. *What are the effects of caloric?*

Heat which, in proportion to its intensity, acts variously upon all bodies, causing *expansion, fusion, evaporation, decomposition, &c.*

75. *Why is caloric called a repulsive agent?*

Because its chief effects are to *expand, fuse, evaporate,* or *decompose* the substances upon which it acts.

76. *What is an attractive agent, in contradistinction to a repulsive agent?*

Chemical attraction, or affinity, is an attractive agent—as when bodies seek of their own natures to unite and form some new body.

77. *When is a body said to be hot?*

When it holds so much *caloric* that it diffuses heat to surrounding objects.

78. *When is a body said to be cold?*

When it holds less *caloric* than surrounding objects, and absorbs heat from them.

79. *How may caloric be excited to develope heat?*

By any means which cause agitation, or produce an active change in the condition of bodies. Thus friction, percussion, sudden condensation or expansion, chemical combination, and electrical discharges, all develope *heat.*

80. *Why do "burning glasses" appear to set fire to combustible substances?*

Because they gather into one point, or *focus,* several rays of *caloric* as they are travelling from the sun, and the accumulation of caloric developes that intensity of *heat* which constitutes *fire.*

81. *What is a focus?*

In optics, it is the point or centre at which, or around which, divergent rays are brought into the closest possible union.

Yet man is born to trouble, as the sparks fly upward.—I would seek unto God, and unto God would I commit my cause."—JOB V.

82. *What is fire?*

It is a violent chemical action attending the combustion of the ingredients of *fuel* with the *oxygen* of the air.

83. *What are the properties of fire?*

It imparts heat, which has the effect of expanding both fluids and solids.

It cannot exist without the presence of combustible materials.

It has a tendency to diffuse itself in every direction.

It cannot exist without oxygen or atmospheric air.

84. *What elements take part in the maintenance of a fire?*

Hydrogen, carbon, and oxygen. Hydrogen and carbon exist in the *fuel*, and oxygen is supplied by the *air*.

85. *How does the combustion of a fire begin?*

A match made of phosphorous and sulphur (highly inflammable substances) is drawn over a piece of sand-paper; the *friction* of the match induces the presence of *caloric*, which developes *heat*, and ignites the match, the burning of which is sustained by the *oxygen* of the air. The flame is then applied to paper or wood, and the heat of the flame is sufficient to drive out *hydrogen gas*, which unites with the *oxygen* of the air, and burns, imparting greater heat to the *carbon* of the coals, which assumes the form of carbonic acid gas by union with *oxygen*, and in a little while all the conditions of *combustion* are established.

86. *What are the properties of heat?*

It may exist without *fire* or *light*.

It is not sensible to *vision*.

It makes an impression upon our *feelings*.

It acts powerfully upon *all bodies*.

It has no *weight*.

It attends, or is connected with, *all the operations of nature*.

It radiates from *all bodies* in straight lines, and in all *directions*.

It strikes most powerfully in *direct lines*.

Its rays may be collected into a *focus*, just as the rays of the sun.

It may be *reflected* from a polished surface.

It is more easily *conducted* by some substances than by others.

"For my days are consumed like smoke, and my bones are burned as an hearth."—PSALM CII.

87. *What is animal heat?*

Animal heat is derived from the slow combustion of *carbon* in the blood of animals with the *oxygen* of the air which the animals breathe.

88. *What is latent heat?*

Latent heat (or more properly *latent caloric*) is that which exists, in some degree, in all *bodies*, though it may be imperceptible to the *senses*.

89. *Is there latent caloric in ice, snow, water, marble, &c?*

Yes; there is some amount of *caloric* in all substances.

[A blacksmith may hammer a small piece of iron until it becomes *red hot*. With this he may light a match, and *kindle the fire of his forge*. The iron has become more dense by the hammering, and it cannot again be heated to the same degree by similar means, until it has been exposed *in fire*, to *a red heat*. Is it not possible that, by hammering, the particles of iron have been driven closer together, and *the latent heat* driven out? No further hammering will force the atoms nearer, and therefore no further heat can be developed. But when the iron has *again absorbed caloric*, by being plunged in a fire, it is again charged with latent heat. Indians produce *sparks* by rubbing together *two pieces of wood*. Two pieces of ice may be rubbed together until sufficient warmth is developed to *melt them both*. The axles of railway carriages frequently become *red hot* from *friction*.]

90. *Have vegetables heat?*

Yes; whenever oxygen combines with carbon to form carbonic acid gas, an extrication of heat takes place, however minute the amount. Such a combination occurs much more extensively during the germination of seeds and the impregnation of flowers, than at any other time. In the germination of barley heaped in rooms, previous to being converted into malt, it is well known that a *considerable amount of heat is developed.*

91. *Has any investigation of this subject ever been carefully made?*

Yes. Lamarck, Senebier, and De Candolle, found the flowers of the *Arum Maculatum*, between three and seven o'clock in the afternoon, as much as 7 deg. Reaum. warmer than the external air. Schultz found a difference of 4 deg. to 5 deg. between the heat of the spathe of the *Canadian pinnatifolium* and the surrounding air, at six to seven o'clock p.m. Other observations have established differences of as much as 30 deg. between the temperature of the spathe of the *Arum cordifolium*, and that of the surrounding atmosphere.

92. *Have plants sometimes a temperature lower than that of the surrounding air?*

Yes. It has not only been found that under particular circumstances the heat of certain parts of plants is elevated to a very remarkable degree, but that, under nearly all circumstances, they have a temperature different from that of the external air, being *warmer in winter, and cooler in summer.*

CHAPTER VI.

93. *How many kinds of combustion are there?*

There are *three*, viz., slow oxydation, *when little or no light is evolved;* a more rapid combination, *when the heat is so great as to become luminous;* and a still more energetic action, *when it bursts into flame.*

94. *Why does phosphorous look luminous?*

Because it is undergoing slow *combustion.*

95. *Why do decayed wood, and putrifying fish, look luminous?*

Because they are undergoing slow *combustion.* In these cases the heat and light evolved are at no one time very considerable. But the *total amount* of *heat*, and probably of *light*, generated through the lengthy period of this slow oxydation, *amounts to exactly the same as would be evolved during the most rapid combustion of the same substances.*

96. *What is flame?*

It is gaseous matter burning at a *very high temperature.*

97. *Why, when we put fresh coals upon a fire, do we hear the gas escaping from the coals without taking fire?*

Because, the fire being slow, the temperature is not high enough to ignite the gas.

98. *What is the gas which escapes from the coals?*

Carburetted hydrogen.

99. *Why, if we light a piece of paper, and lay it where the gas is escaping from the coals, will it burst into flame?*

Because the lighted paper gives a *heat sufficient* to ignite the gas; and because also hydrogen requires the contact of *flame* to ignite it.

100. *Why, when the coals have become heated, will the hydrogen burst into flame?*

Because the *carbon* of the coals, and the *oxygen* of the air, have begun to combine, and have greatly increased the *heat*, and have produced a rapid combustion, *so nearly allied to flame*, that it *ignites the hydrogen.*

101. *What temperature is required to produce flame?*

That depends upon the nature of the combustible you desire to burn. Finely divided phosphorous and phosphorated hydrogen will take fire at a temperature of 60 deg. or 70 deg.; solid phosphorous at 140 deg.; sulphur at 500 deg.; hydrogen and carbonic oxide at 1,000 deg. (red heat); coal gas, ether, turpentine, alcohol, tallow, and wood, at about 2,000 deg. (incipient white heat). When once inflamed they will *continue to burn*, and will maintain a very high temperature.

102. *What is smoke?*

Smoke consists of small particles of *carbon* of *hydrogen gas*, and *other volatile matters*, which are driven off by heat and carried up the chimney.

103. *Is it not a waste of fuel to allow this matter to escape?*

It is, as it might all be burnt up by better management.

104. *How may the waste be avoided?*

By putting on only a little coals at a time, so that the heat of the fire shall be sufficient to consume these volatile matters as they escape.

105. *Why is there so little smoke when the fire is red?*

Because the *hydrogen* and the *volatile* parts of the *coal* have already been driven off and consumed, and the combustion that continues is principally caused by the *carbon* of the coals, and the oxygen of the air.

106. *Will carbon, burnt in oxygen, produce flame and smoke?*

It burns brightly, but it produces neither flame nor smoke.

107. *Why do not charcoal and coke fires give flame?*

Because the *hydrogen* has been driven off by the processes by which charcoal and coke are made.

108. *What is a conductor of heat?*

A conductor of heat is any substance through which heat is *readily transmitted.*

109. *What is a non-conductor of heat?*

A non-conductor is any substance through which heat will *not* pass readily.

110. *Name a few good conductors.*

Gold, silver, copper, platinum, iron, zinc, tin, stone, *and all dense solid bodies.*

111. *Name a few non-conductors.*

Fur, wool, down, wood, cotton, paper, and *all substances of a spongy or porous texture.*

112. *How is heat transmitted from one body to another?*

By Conduction, Radiation, Reflection, Absorption and Convection.

113. *What is the Conduction of heat?*

It is the communication of heat from one body to another *by contact.* If I lay a penny piece upon the hob, it becomes hot by *conduction*

114. *What is the Radiation of heat?*

The transmission of heat by a *series of rays.* If I hold my hand before the fire, the rays of heat fall upon it, and *my hand receives the heat through radiation.*

115. *What is the Reflection of heat?*

The reflection of heat is the *throwing back* of its rays towards the direction whence they came. In a Dutch oven the rays of heat pass from the fire to the oven, and are *reflected* back again by *the bright surface of the tin.* There is, therefore, considerable economy of heat in ovens, and other cooking utensils constructed upon this plan.

116. *What is the Absorption of heat?*

The absorption of heat is the taking of it up by the body to which it is transmitted or conducted. Heat was conveyed to my hand by *radiation,* and *taken up* by my hand by *absorption*

117. *What is the Convection of heat?*

The convection of heat is the transmission of it *through* a body or a number of bodies, or particles of bodies, by those substances which *first received it;* as when hot water rises from the bottom of a kettle and imparts heat to the cold water lying above it.

CHAPTER VII.

118. *Why does not a piece of wood which is burning at one end, feel hot at the other end?*

Because wood is *a bad conductor of heat.*

119. *Why is wood a bad conductor of heat?*

Because the arrangement of the particles of which it is composed does not favour the transmission of *caloric.*

120. *Why do some articles of clothing feel cold, and others warm?*

Because some are bad conductors of heat, *and do not draw off much of the warmth of our bodies;* while others are *better conductors,* and *take up a larger portion of our warmth.*

"The fining pot is for silver, and the furnace for gold: but the Lord trieth the hearts."—PROVERBS XVII.

121. *Which feels the warmer, the conductor or non-conductor?*

The non-conductor, as it does not readily *absorb* the warmth of our bodies.

122. *What substances are the best conductors of heat?*

Gold, silver, copper, and most substances of close and hard formation, &c.

123. *What substances are the worst conductors of heat?*

Fur, eider down, feathers, raw silk, wood, lamp-black, cotton, soot, charcoal, &c.

124. *Why has the toasting-fork a wooden handle?*

Because wood is not *so good a conductor* as metal, therefore the wood prevents the heat from being transmitted *by conduction* to our hands.

125. *Why has the coffee-pot a wooden handle?*

Because the metal of the coffee-pot would otherwise *conduct the heat to the hand;* but wood, *being a bad conductor,* prevents it.

126. *Why does hot water in a metal jug feel hotter than in an earthenware one?*

Because metal, being a good conductor, *readily delivers heat to the hand;* but *earthenware, being an indifferent conductor,* parts with the heat slowly.

127. *How can we ascertain that wood prevents the conduction of heat to the hand?*

By passing the top of the finger along the wooden handle of the coffee-pot, until it reaches the point where the wood meets the metal. The wooden handle will be found to be *cool,* but the metal will feel *very hot.*

128. *Of what use are kettle-holders?*

Being made of *bad conductors,* such as wood, paper, or woollen cloth, they will not readily *conduct* the heat from the kettle to the hand.

129. *Will a kettle-holder, being a bad conductor, sometimes conduct heat to the hand?*

Yes. But so slowly that the hand will not *feel the inconvenience of too much heat.*

130. *Why does hot metal feel hotter than heated wool, though they may both be of the same degree of temperature?*

Because metal gives out heat *more rapidly than wool,* bv which it is made *more perceptible to our feelings.*

131. *Which would become cold first—the metal or the wool?*

The *wool,* because, although the metal conducts heat more rapidly, to a substance in contact with it, it does not *radiate heat* as well as a *black and rough substance.*

132. *Why do iron articles feel intensely cold in winter?*

Because iron is one of the best conductors, and draws off heat from the hand very rapidly.

133. *What is the cause of the sensation called cold?*

When we feel cold, heat is being *drawn off from our bodies.*

134. *What is the cause of the sensation called heat?*

When we feel hot, our bodies are *absorbing heat* from external causes.

[The condition here implied is that of health, and of ordinary circumstances. A person in a condition of fever, suffering from intense heat arising from a diseased state of the blood, could not be said to be *absorbing heat.* Nor could such a description apply to a person who, by a very rapid walk, has raised the temperature of his body considerably above its natural state, by the *internal combustion* which has already been described. A person feeling hot in bed, from excessive clothes, feels hot from the *development of heat internally,* which is not *conducted away* with sufficient rapidity to maintain the natural temperature of the body.]

135. *If a person, sitting before a fire-place, without a fire, were to set one foot upon a rug, and the other upon the stone hearth, which would feel the colder?*

The foot on the stone, because stone is a good conductor, and would *conduct the warmth of the foot away from it.*

136. *What does the hearth-stone do with the heat that it receives?*

It delivers it to the surrounding air, and to any other bodies with which it may be *in contact*—and as it parts with heat, *it takes up more from any body hotter than itself.*

137. *When there is no fire in a room, what is the relative temperature of the various things in the room?*

They are all of the same temperature.

138. *If all the articles in the room are of the same temperature, why do some feel colder than others?*

Because they differ in their relative powers of *conduction.* Those that are the best conductors feel coldest, as they convey away the heat of the hand most rapidly.

[If you lay your hand upon the *woollen table cover*, or upon the *sleeve of your coat* or mantle, it will feel *neither warm nor cold*, under ordinary circumstances. But if you raise your hand from the table cover, or coat, and lay it on the marble mantel piece, the mantel-piece will feel *cold.* If now you return your hand from the mantel-piece to the table cover or coat, *a sensation of warmth will become distinctly perceptible.* This will afford a good conception of the relative *powers of conduction of wool and marble.*]

139. *How long does a substance feel cold or hot to the touch?*

Until it has brought the part touching it to the same temperature as itself.

140. *When do substances feel neither hot nor cold?*

When they are of the same temperature as our bodies.

141. *Why, under these circumstances, do they feel neither hot nor cold?*

Because they neither take heat from, nor supply it to, the body.

142. *Which would feel the warmer, when the fire was lighted, the hearth-rug or the hearth-stone?*

The hearth-stone, because it is a *good conductor*, and would not only *receive heat* readily, but would *part with it as freely* (thereby making its heat *perceptible*). But the hearth-rug, *being a bad conductor*, would part with its heat very slowly, and it would therefore be *less perceptible*.

143. *Would the hearth-stone feel hotter than the hearth-rug though both were of the same temperature?*

It would feel *hotter than the hearth-rug*, because it would part with its heat so rapidly that it would be the *more perceptible*.

144. *But if the hearth-stone and the hearth-rug were both colder than the hand, which would feel the colder of the two?*

Then the hearth-stone would feel the colder, because, *being a good conductor*, it would *take heat* from the hand more freely than the hearth-rug, which is a *bad conductor*.

145. *Why would the hearth-stone feel comparatively hotter in the one case, and colder in the other?*

Because, *being a good conductor*, it would conduct heat rapidly *to* the hand when hot, and take heat rapidly *from* the hand when cold.

CHAPTER VIII.

146. *Which are the better conductors of heat, fluids or solids?*

Generally speaking, *solids*, especially those of them that are dense in their substance.

147. *Why are dense substances the best conductors of heat?*

Because the heat more readily travels from particle to particle until it pervades the mass.

148. *Why are fluids bad conductors of heat?*

Because of the want of *density* in their bodies; and because a portion of the imbibed heat always passes off from fluids by *evaporation*.

"He casteth forth his ice like morsels: who can stand before his word."—
PSALM CXLVII.

149. *Why are woollen fabrics bad conductors of heat?*

Because there is a considerable amount of *air* occupying the spaces of the texture.

150. *Is air a good or a bad conductor?*

Air is a *bad conductor*, and it chiefly transmits heat, as water does, by *convection*.

151. *Is water a good or a bad conductor?*

Water is an indifferent conductor, but it is a *better conductor than air*.

152. *Why, when we place our hands in water, which may be of the same temperature as the air, does the water feel some degrees colder?*

Because water, *being a better conductor than air*, takes up the warmth of the hand *more rapidly*.

153. *Why, when we take our hands out of water do they feel warmer?*

Because the air does not abstract the heat of the hand so rapidly as the water did, and the change in the degree of rapidity with which the heat is abstracted *produces a sensation of increased warmth*.

154. *Why do we see blocks of ice wrapped in thick flannel in summer time?*

Because the flannel, being a non-conductor, prevents the *external heat* from *dissolving the ice*.

[Flannel wrapped around a *warm* body *keeps in its heat;* and wrapped around a *cold* body, prevents heat from *passing into it*.]

155. *How do we know that air is not a good conductor of heat?*

Because, *in still air*, heat would travel to a given point much more rapidly, and in greater intensity, through even an indifferent *solid conductor*, than it would through the *air*.

156. *How do we know that water is not a good conductor of heat?*

"As snow in summer, and as rain in harvest; so honour is not seemly for a fool."—PROV. XXVI.

Because in a deep vessel containing *ice*, and with heat applied at the top, some portion of the water may be made to boil *before the ice, which lies a little under the surface, is melted.*

157. *Why would you apply the heat at the top, in this experiment?*

Because in heating water it *expands and rises.* The boiling of water is caused by the heated water *ascending from the bottom,* and the colder water descending to occupy its place. If the heat were not applied at the top, it would be distributed quickly by *convection,* but not by *conduction.*

158. *Why are bottles of hot water, used as feet-warmers, wrapped in flannel?*

Because the flannel, *being a bad conductor,* allows the heat to *pass only gently* from the bottle, and preserves the warmth for a *much longer time.*

159. *Why are hot rolls sent out by the bakers, wrapped up in flannel?*

Because the flannel, *being a bad conductor,* does not *carry off rapidly the heat of the rolls.*

160. *Why is it said that snow keeps the earth warm?*

Because snow is a *bad conductor,* and prevents the frosty air from *depriving the earth of its warmth.*

161. *Why are snow huts which the Esquimaux build found to be warm?*

Because snow, *being a bad conductor,* keeps in *the internal heat of the dwelling,* and prevents the *cold outer air from taking away its warmth.*

162. *Why is snow, being composed of congealed water (and water being a better conductor than air), so good a non-conductor?*

Because in the process of congealation it is frozen into crystaline forms, which, being collected into a mass, form a woolly body, thus proving the truthfulness of the Bible simile, which says, God "giveth snow like wool."

CRYSTALS OF SNOW, AS SEEN THROUGH A MICROSCOPE. FIG. 1.

163. *Why does it frequently feel warmer after a frost has set in?*

Because, in the act of congelation a great deal of heat is given out, and *taken up by the air*, and thus *the severity of the cold is in some degree moderated.*

164. *Why is it frequently colder when a thaw takes place?*

Because, in the process of thawing, a certain amount of heat is *withdrawn from the air*, and enters the thawed ice.

165. *What benefit results from these provisions of Nature?*

They moderate both the *severity of frosts*, and *the rapidity of thaws*, which, in changeable climates, would be seriously detrimental to *life*, and to *vegetation.*

166. *Why are furs and woollens worn in the winter?*

Because, being non-conductors, they prevent the warmth of the body from being *taken up by the cold air.*

167. *Why are the skins of animals usually covered with fur, hair, wool, or feathers?*

Because their coverings, being *non-conductors of heat,* preserve the warmth of the bodies of the animals.

168. *How is the greater warmth of animals provided for in the winter?*

It is observed that, as winter approaches, there comes a short woolly or downy growth, which, *adding to the non-conducting property of their coats,* confines their animal warmth.

[In small birds during winter, let the external colour of the feathers be what it may, there will be found a kind of *black* down next their bodies. Black is the *warmest colour,* and the purpose here is to *keep in the heat,* arising from the respiration of the animal.]

169. *How is warmth provided for in animals that have no such coats?*

They are furnished with a layer of *fat,* which lies underneath the skin. Fat consists chiefly of *carbon,* and is a *non-conductor.*

170. *Why are summer breezes said to be cool?*

Because, as they pass over the heated surface of the body, they bear away a part of its heat.

171. *Why is a still summer air said to be sultry?*

Because, being heated by the sun's rays, *and being a bad conductor,* it does not relieve the body by *carrying off its heat.*

172. *Why does fanning the face make it feel cooler?*

Because, by inducing currents of air to pass over the face, a part of the excessive heat is taken up *and carried away.*

173. *Why does perspiration cool the body?*

Because it takes up a part of the heat, and, evaporating, *carries it into the air.*

174. *Why does blowing upon hot tea cool it?*

Because it directs currents of air over the surface of the tea, and these currents take up a part of the heat *and bear it away.*

175. *Why does air in motion feel cooler than air that is*

Because each wave of air *carries away a certain portion of heat* and being followed by another portion of air, *a further amount of heat is borne away.*

176. *Is the atmosphere ever as hot as the human body?*

Not in this country. On the hottest day it is 10 or 12 deg. *cooler than the temperature of our bodies.*

177. *What is the highest degree of artificial heat which man has been known to bear?*

A man may be surrounded with air raised to the temperature of 300 deg. (the boiling point being 212), and yet not have the heat of his body raised more than two or three degrees above its natural temperature of from 97 deg. to 100 deg.

178. *Why may man endure this degree of heat for a short time without injury?*

Because the skin, and the vessels of fat that lie underneath it, are bad conductors of heat.

And because perspiration passing from the skin and evaporating, would *bear the heat away* as fast as it was received.

Because, also, the vital principle (life) exercises a mysterious influence in the preservation of living bodies from physical influences.

179. *Is the air ever hot enough, in any part of the world, to destroy life?*

Yes. The hot winds of the Arabian deserts, which are called *simooms*, scatter death and desolation in their track, withering trees and shrubs, and burying them under waves of hot sand. When camels see the approach of a simoom they rush to the nearest tree or bush, or to some projecting rock, where they place their heads in an opposite direction to that from which the wind blows, and endeavour to escape its terrible violence. The traveller throws himself on the ground on the lee side of the camel, and screens his head from the fiery blast within the folds of his robe. But frequently both man and beast *fall a prey to the terrible simoom.*

180. *Why are these hot winds so terrible in their effects?*

Because, being in motion they search their way to every part of the body, and passing over it *leave some portion of their heat behind,* which is again followed by *additional heat from every fresh blast of wind.*

CHAPTER IX.

181. *What is Radiation?*

The radiation of heat is a *motion of the particles,* in a series of rays, diverging in every direction from a heated body.

182. *What is this phenomena of Radiation understood to arise from?*

From a strongly repulsive power, possessed by particles of heat, by which they are excited to recede from each other with great velocity.

183. *What is the greatest source of Radiation?*

The sun, which sends forth rays of *both light and heat* in all directions.

184. *When does a body radiate heat?*

When it is surrounded by a medium which is *a bad conductor*

185. *When we stand before a fire, does the heat reach us by conduction or by radiation?*

By radiation.

186. *What becomes of the heat that is radiated from one body to another?*

It is either *absorbed* by those bodies, or transmitted through them and passed to other bodies by *conduction,* or diffused by *convection,* or returned by *reflection.*

187. *How do we know that heat is diffused by radiation?*

If we set a metal plate (or any other body, though metal is best for the experiment) before the fire, *rays of heat will fall upon it* If we turn the plate at a slight angle, and place another object in a line with it, we shall find that the plate will *reflect the rays it has received by radiation*, on to the object so placed; but if we place an object *between the fire and the plate*, we shall find that the rays of heat *will be intercepted*, and that the latter can no longer *reflect heat*.

188. *Does the agitation of the air interfere with the direction of rays of heat?*

It has been found that the agitation of the air does *not* affect the direction of rays of heat.

189. *Why, then, if a current of air passes through a space across which heat is radiating, does the air become warmer?*

Because it takes up *some portion of the heat*, but it does not alter the direction of the rays.

[This is clearly illustrated by reference to *rays of light* which are seen under many circumstances. But they are never bent, moved, nor in any way affected by the wind.]

190. *Why will not a current of air disturb the rays of heat, just as it would a spider's web, or threads of silk?*

Because heat is an *imponderable* agent, that is, something which cannot be acted upon by the ordinary physical agencies. It has *no weight*, presents no *substantial body*, and is, in these latter respects, similar to *light and electricity*.

191. *What other sources of radiation of heat are there besides the sun and the fire?*

The *earth*, and all *minor bodies*, are, in some degree, *radiators of heat*.

192. *What substances are the best radiators?*

All *rough* and *dark* coloured substances and surfaces are the *best radiators of heat*.

193. *What substances are the worst radiators of heat?*

All *smooth, bright*, and *light coloured* surfaces are *bad radiators of heat*.

[Dr. Stark, of Edinburgh, has proved, by a series of experiments, the influence which the *colours* of bodies have upon the *velocity of radiation*. He surrounded the bulb of a thermometer successively with equal weights of *black*, *red*, and *white* wool, and placed it in a glass tube, which was heated to the temperature of 180 deg. by immersion in hot water. The tube was then cooled down to 50 deg. by immersion in cold water; the *black* cooled in 21 minutes, the *red* in 26 minutes, and the *white* in 27 minutes.]

194. *If you wished to keep water hot for a long time, should you put it into a bright metal jug, or into a dark earthenware one?*

You should put it into a *bright metal* jug, because, *being a bad radiator*, it would not part readily with the heat of the water.

195. *Why would not the dark earthenware jug keep the water hot as long as the bright metal one?*

Because the particles of earthenware being rough, and of dark colour, *they radiate heat freely*, and the water would thereby be quickly cooled.

CHAPTER X.

196. *But if (as stated in the Lessons upon Conduction) metal is a better conductor of heat than stone or earthenware, why does not the metal jug conduct away the heat of the water sooner than the earthenware jug?*

It would do so, *if it were in contact with another conductor;* but, being surrounded by air, *which is a bad conductor*, the heat must pass off *by radiation*, and as bright metal surfaces are bad radiators, the metal jug would retain the heat of the water *longer than the earthenware one.*

197. *Supposing a red-hot cannon ball to be suspended by a chain from the ceiling of a room, how would its heat escape?*

Almost entirely by *radiation*. But if you were to rest upon the ball a cold bar of iron, a part of the heat would be drawn off by *conduction*. Warm air would rise from around the ball, and, moving upwards, would distribute some of the heat by *convection*

And some of its rays, falling upon a mirror, or any other bright surface, might be diffused by *reflection*.

198. *Do some substances absorb heat?*

Yes; those substances which are *the best radiators* are also *the best absorbers* of heat.

199. *Why does scratching a bright metal surface increase its power of radiation?*

Because every irregularity of the surface acts as a point of radiation, or *an outlet* by which the heat escapes.

200. *Why does a bright metal tea-pot produce better tea than a brown or black earthenware one?*

Because bright metal *radiates but little heat*, therefore the water is kept hot much longer, *and the strength of the tea is extracted by the heat.*

201. *But if the earthenware tea-pot were set by the fire, why would it then make the best tea?*

Because the dark earthenware tea-pot is a good *absorber of heat*, and the heat it would *absorb* from the fire would more than counterbalance the loss by *radiation*.

202. *How would the bright metal tea-pot answer if set upon the hob by the fire?*

The bright metal tea-pot would probably *absorb less heat* than it would radiate. Therefore it would not answer so well, *being set upon the hob*, as the earthenware tea-pot.

203. *Why should dish covers be plain in form, and have bright surfaces?*

Because, being bright and smooth, they will not allow heat to escape *by radiation*.

204. *Why should the bottoms and back parts of kettles and saucepans be allowed to remain black?*

Because a *thin* coating of soot acts as a *good absorber of heat*, and overcomes the *non-absorbing* quality of the *bright surface*.

205. *But why should soot be prevented from accumulating in flakes at the bottom and sides of kettles and saucepans?*

Because, although soot is a *good absorber* of heat, it is a *very bad conductor;* an accumulation of it, therefore, would cause a waste of fuel, by *retarding the effects of heat.*

206. *Why should the lids and fronts of kettles and saucepans be kept bright?*

Because bright metal *will not radiate heat;* therefore, the heat which is taken up readily through the *absorbing* and *conducting* power of the bottom of the vessel, is kept in and economised by the *non-radiating* property of the bright top and front.

207. *Does cold radiate as well as heat?*

It was once thought that *cold radiated* as well as *heat.* But a mass of ice can only be said to radiate cold, *by its radiating heat in less abundance than that which is emitted from other bodies surrounding it.* It is, therefore, *incorrect* to speak of the *radiation of cold.*

CHAPTER XI.

208. *Why, if you hold a piece of looking-glass at an angle towards the sun, will light fall upon an object opposite to the looking-glass?*

Because the rays of the sun are *reflected* by the looking-glass.

209. *Why, when we stand before a mirror, do we see our features therein?*

Because the rays of light that fall upon us are *reflected* upon the bright surface of the mirror.

210. *Why, if a plate of bright metal were held sideways before a fire, would heat fall upon an object opposite to the plate?*

"But the wise answered saying, Not so; lest there be not enough for us and you: but go ye rather to them that sell, and buy for yourselves."—MATT. XXV.

Because rays of heat may be *reflected* in the same manner as the rays of light.

211. *Why would not the same effect arise if the plate were of a black or dark substance?*

Because black and dark substances are not *good reflectors of heat.*

212. *What are the best reflectors of heat?*

Smooth, light-coloured, and highly polished surfaces, especially those of *metal.*

213. *Why does meat become cooked more thoroughly and quickly when a tin screen is placed before the fire?*

Because the bright tin reflects the rays of heat back again to the meat.

214. *Why is reflected heat less intense than the primary heat?*

Because it is impossible to collect all the rays, and also because a portion of the caloric, imparting heat to the rays, is absorbed by the air, and by the various other bodies with which the rays come in contact.

215. *Can heat be reflected in any great degree of intensity?*

Yes; to such a degree that inflammable matters may be ignited by it. If a cannon ball be made red hot, and then be placed in an iron stand between two bright reflectors, inflammable materials, placed in a proper position to catch the reflected rays, *will ignite from the heat.*

There is a curious and an exceptional fact with reference to *reflected* heat, for which we confess that we are unable to give "*The Reason Why.*" It is found that snow, which lies near the trunks of trees or the base of upright stones, melts before that which is at a distance from them, though the sun may shine equally upon both. If a blackened card is placed upon ice or snow under the sun's rays, the frozen body underneath it will be thawed before that which surrounds it. But if we *reflect* the sun's rays from a metal surface, the result is *directly contrary*—the exposed snow is the first to melt, leaving the card standing as upon a pyramid. Snow *melts* under heat which is *reflected* from the trees or stones while it withstands the effect of the *direct solar rays.* In passing through a cemetery this winter (1857), when the snow lay deep, we were struck with the circumstance that the snow in front of the head-stones facing the sun was completely dissolved, and, in nearly every instance, the space on which the snow had melted assumed a coffin-like shape. This forced itself so much upon our attention that we remained some time to endeavour to analyse the phenomena; and it was not until we remembered the curious effect of *reflected heat* that we could account for it. It is obvious that the rays falling from the upper part of the head-stone on to the *foot* of the grave would be less powerful than those that radiated from the *centre* of the stone to the centre of the grave. Hence it was that the heat dissolved at the foot of the grave only a narrow piece of snow, which widened towards the centre, and narrowed again as it approached the foot of the head-stone, where the lines of radiation would naturally decrease. Such a phenomena would prove sufficient to raise superstition in untutored minds.

216. *Are good reflectors of heat also good absorbers?*

No; for reflectors at once *send back* the heat which they receive, while absorbers *retain it.* It is obvious, therefore, that *reflectors* cannot be good *absorbers.*

217. *How do fire-screens contribute to keep rooms cool?*

Because they turn away from the persons in the room rays of heat which would otherwise make the warmth excessive.

218. *Why are white and light articles of clothing cool?*

Because they *reflect* the rays of heat.

[White, as a *colour*, is also a bad *absorber* and *conductor.*]

219. *Why is the air often found excessively hot in chalk districts?*

Because the soil *reflects* upon objects near to it the heat of the solar rays.

220. *How does the heat of the sun's rays ultimately become diffused?*

It is first *absorbed* by the earth. Generally speaking, the earth *absorbs* heat by day, and *radiates* it by night. In this way an equilibrium of temperature is maintained, which we should not otherwise have the advantage of.

221. *Does not the air derive its heat directly from the sun's rays?*

Only partially. It is estimated that the air absorbs only *one-third* of the caloric of the sun's rays—that is to say, that a ray of solar heat, entering our atmosphere at its most attenuated limit (a height supposed to be about *fifty miles*), would, in passing through the atmosphere to the earth, part with only one-third of its calorific element.

222. *What becomes of the remaining two-thirds of the solar heat?*

They are ***absorbed*** chiefly by the ***earth***, the great medium of calorific ***absorption;*** but some portions are taken up by ***living things***, both animal and vegetable. When the ***rays of heat*** strike upon the earth's surface, they are passed from particle to particle into the interior of the earth's crust. Other portions are distributed through the air and water by ***convection***, and a third portion is thrown back into space by ***radiation***. These latter phenomena will be duly explained as we proceed.

223. *How do we know that heat is absorbed, and conducted into the internal earth?*

It is found that there is a given depth beneath the surface of the globe at which an equal temperature prevails. The depth increases as we travel south or north from the equator, and corresponds with the shape of the earth's surface, ***sinking under the valleys, and rising under the hills.***

224. *Why may we not understand that this internal heat of the earth arises, as has been supposed by many philosophers, from internal combustion?*

Because recent investigations have thrown considerable and satisfactory light upon the subject. It has been ascertained that the internal temperature of the earth ***increases*** to a certain depth, ***one degree in every fifty feet.*** But that below that depth the temperature ***begins to decline***, and continues to do so with every increase of depth.

225. *Do plants absorb heat?*

Yes. They both ***absorb*** and ***radiate*** heat, under varying circumstances. The majestic tree, the meek flower, the unpretending grass, all perform a part in the grand alchemy of nature.

"Consider the lilies of the field, how they grow; they toil not, neither do they spin.

When we gaze upon a rose it is not its beauty alone that should impress us: every moment of that flower's life is devoted to the fulfilment of its part in the grand scheme of the universe. It decomposes the rays of solar light, and sends the red rays only to our eyes. It absorbs or radiates heat, according to the temperature of the aerial mantle that wraps alike the flower and the man It distils the gaseous vapours, and restores to man the vital air on which he lives. It takes into its own substance, and incorporates with its own frame, the carbon and the hydrogen of which man has no immediate need. It drinks the dew-drop or the rain drop, and gives forth its sweet odour as a thanksgiving. And when it dies, it preaches eloquently to beauty, pointing to the end that is to come!

CHAPTER XII.

226. *How do we know that plants operate upon the solar and atmospheric heat?*

A delicate thermometer, placed among the leaves and petals of flowers, will at once establish the fact, not only that flowers and plants have a temperature differing from that of the external air, but that the temperature varies in different plants according to the hypothetical, or supposed requirements, of their existences and conditions.

227. *What is the chief cause of variation in the temperature of flowers?*

It is generally supposed that their temperature is affected by their *colours.*

228. *Why is it supposed that the colour of a flower influences its temperature?*

Because it is found by experiment that the *colours* of bodies bear an important relation to their properties respecting *heat,* and hold some analogy to the relation of *colours* to *light.*

[If when the ground is covered with snow, pieces of woollen cloth, of equal size and thickness, and differing only in colour, are laid upon the surface of the snow, near to each other, it will be found that the relation of *colour* to temperature will be as follows:—In a few hours the *black* cloth will have dissolved so much of the snow beneath it, as to sink deep below the surface: the *blue* will have proved nearly as warm as the black; the *brown* will have dissolved less of the snow: the *red* less than the brown; and the *white* the least or none at all. Similar experiments may be tried with reference to the *condensation of dew*, &c. And it will be uniformly found that the *colour* of a body materially affects its powers of *absorption* and of *radiation*.]

229. *Why do we know that these effects are not the result of light?*

Because they would occur, in just the same order, in the absence of light.

230. *Why are dark coloured dresses usually worn in winter, and light in summer?*

Because black *absorbs* heat, and therefore becomes warm; while *light colours* do *not* absorb heat in the same degree, and therefore they remain cool.

231. *Why do iron articles, even when near fire, usually feel cool?*

Because they are bad absorbers, and do not take up heat freely, unless they are *in contact* with a hot body.

232. *How is heat diffused through the atmosphere?*

By *convection.* The warmth radiating from the surface of the earth warms the air in contact with it; the air expands, and becoming lighter, flies upwards, bearing with it the caloric which it holds, and diffusing it in its course.

233. *How do the waters of the ocean become heated?*

Chiefly by *convection.* Nearly all the heat which the sun sheds upon the ocean is borne away from its surface by evaporation, or is radiated back into the atmosphere. But the ocean gathers its heat by *convection* from the earth. It girdles the shores of tropical lands where, being warmed to a high degree of temperature, it sets across the Atlantic from the Gulf of Mexico, and exercises an important influence upon the temperature of our latitude.

234. *What is the cause of winds?*

Currents of air, and winds, are the result of *convection.* The air, heated by the high temperature of the tropics, *ascends*, while the colder air of the temperate and the frigid zones *blows towards the equator* to supply its place.

"Give unto the Lord the glory due unto his name; worship the Lord in the beauty of holiness."—PSALM xxix.

235. *What is the cause of sea breezes?*

Sea breezes are also the result of *convection.* The land, under the heat of the day's sunshine, becomes of a high temperature, and the expanded air on its surface *flies away towards the ocean.* As the sun goes down, the earth cools again, and the air *flies back* to find its equilibrium.

Many countries by the sea are subjected to these periodical breezes, known as either "land" or "sea breezes," according to their direction. About eight o'clock in the morning an aerial current begins to flow from the sea towards the land, and continues until about three o'clock in the day; then the current takes a reverse direction, flowing from the land to the sea. This it continues to do throughout the night, until the time of sun-rise, when a temporary calm ensues.

236. *Why does a soap bubble ascend in the air?*

Because, being filled with *warm* air, it is *lighter* than the surrounding medium, and therefore ascends.

237. *Why does the bubble fall after it has been in the air some time?*

Because the air contained in it has become cool, and, as it contains carbonic acid gas, it is *heavier* than the air.

238. *What became of the warmth at first contained in the bubble?*

It has been *distributed in the air* through which the bubble passed.

239. *What does this simple illustration of the distribution of warmth explain?*

It explains the law of *convection,* or *heat distribution,* over the surface of the globe.

240. *Why does air ascend the chimney?*

Because, being heated, it becomes *lighter* than the surrounding medium, and therefore flies upwards, through the outlet provided for it.

241. *Why does air fly from the doors and windows towards the fire-place?*

Because, as the warm air flies away, cold air rushes in to occupy its place.

242. *What does this example of the motion of the air in our rooms explain?*

It explains the movement of volumes of air by *convection*, and illustrates the origin of *breezes* and *winds*.

243. *What is the chief effect of this law of convection?*

Under its influence air and water are the great *equalisers of solar heat*, rendering the earth agreeable to living things, and suited to the laws of their existence.

Owing, also, to this law of *convection*, the constituents of the air are equalised. The breath of life, supplied by the purer oxygen of the "sunny south," is diffused in salubrious gales over the wintry climes of the north. And the waters, evaporated from the bosom of the central Atlantic Ocean and the Pacific, are borne across vast continents, and poured down in fertilising showers upon distant lands.

To the educated mind, nothing is too simple to merit attention. To the ignorant, few things are sufficiently attractive to excite curiosity. Knowledge enables us to estimate the varied phenomena that are hourly arising around us, and to see, even in the most trifling effects, illustrations of those great causes and consequences that govern with mighty power the material world. Man, sitting by his fire-side, is enabled to witness the operation of some of nature's grandest laws: *light* and *heat* are around him; *conduction, radiation, reflection, absorption*, and *convection* of heat are all going on before him; little winds are sweeping by his footstool, and warm currents, with miniature clouds folded in their arms, are passing upward before his view. Chemical changes are going on; the solid rock of coal disappears, flying away as an invisible gas. The little "hills are melted," and hard stones have been converted into "fervent heat." Although some of these changes are imperceptible to the *eye*, they are manifest to the educated *mind;* and the pleasures of philosophical observation are as sweet as a poet's dreams.

CHAPTER XIII.

244. *Why will a piece of paper, held three or four inches over the flame of a candle, become scorched?*

"Neither do men light a candle, and put it under a bushel, but on a candlestick; and it giveth light unto all that are in the house."—MATT. V.

Because the hot air and gas produced by the burning of the candle *ascends* rapidly.

245. *Why will a piece of paper held about an inch below the flame of a candle scarcely become warmed?*

Because the heat *ascends;* and only a little of it falls upon the paper, and that by *radiation.*

246. *Why does the lower part of the flame of a candle* (D) *burn of a blue colour?*

Because the *hydrogen* of the tallow, having a stronger affinity for the *oxygen* of the air than *carbon* has, ignites first. Pure hydrogen burns with a bluish flame.

247. *Why does the middle of the flame* (C) *look dark?*

Because it is occupied with gaseous vapours, derived from the tallow, which have not yet *ignited.*

248. *Why does the upper part of the flame* (B) *produce a bright yellow light?*

Because it is in this part of the flame that the *hydrogen* of the candle, and the *oxygen* of the air, combine, and there is just sufficient *carbon* mixed with the *hydrogen* to improve its *illuminating power.*

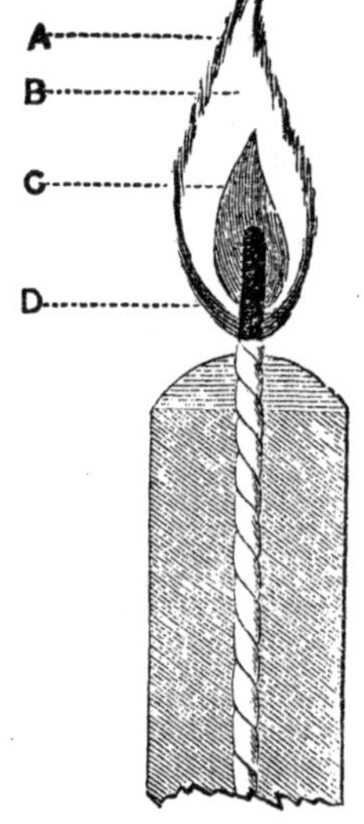

FIG. 2.—DIAGRAM SHOWING THE COMBUSTION OF A CANDLE.

249. *Why is there a fringe of pale light* (A) *around the upper part of the flame?*

Because some of the *carbon* escapes in a state of *incandescence,* and as soon as it reaches the air it combines with *oxygen,* and so forms *carbonic acid gas.*

If any dark body, such as the blade of a knife, be held between the eye and the flame of the candle, so as to shut off the light of the more luminous part, the pale fringe around the flame will be found distinctly perceptible. *Incandesence* means *heated to whiteness.*

250. *Why does the flame terminate in a point?*

Because cold air rushes towards the flame in every direction, and is carried upward. At the point where the flame terminates the cold currents have so *reduced the temperature* that combustion can no longer be sustained.

251. *Why, if you hold anything immediately over the flame, will the flame lengthen?*

Because, by preventing the rapid escape of the heated air, you maintain a temperature which *increases the combustion* at the point of the flame.

252. *Why should persons whose clothes take fire, throw themselves down?*

Because flame spreads most rapidly in an *upward* direction.

253. *Why should persons whose clothes are on fire roll slowly about when they are down?*

Because they thereby *press out* the fire.

254. *Why does pressing a flame or a spark put it out?*

Because it prevents the contact of the flame or spark with the *oxygen* of the air.

Extinguishers put out the flame of candles in the same manner. A person dies from "suffocation" through the absence of oxygen; and it is literally practicable to "*suffocate*" *a fire.*

255. *Why does the wick turn black as it burns?*

Because it consists principally of *carbon.*

256. *Why, when the point of the wick turns out and meets the air, does it exhibit a bright spark?*

Because the *carbon* of the wick comes into immediate contact with the *oxygen* of the air.

257. *Why does holding a candle "upside down" put it out?*

Because the melted grease runs down too rapidly, and at too low a temperature to undergo combustion. It therefore *reduces the heat*, and extinguishes the flame.

258. *Why is it more difficult to blow out the flame of a candle with a cotton wick than one with a rush wick?*

Because the cotton wick imbibes more of the combustible materials, and holds in its loose texture the inflammable gases in a state ready for combustion.

259. *Why does blowing sharply at a candle flame put it out?*

Because the breath drives away the vapour of the grease which, becoming gaseous, supports the flame.

And because too rapid a flow of cold air reduces the temperature below the point at which combustion can be maintained.

260. *Why will a gentle puff of breath, if given speedily after the flame is extinguished, re-kindle it?*

Because the *oxygen* of the air combines with the *carbon* and *hydrogen* that are still escaping from the *heated wick*, and re-lights it.

261. *Why will not a similar puff re-kindle the flame of a rushlight?*

Because its wick retains but *little heat*, and holds a comparatively small amount of combustible matter in a *volatile state*.

262. *Why is a fire, when it is very low, sometimes put out by blowing it?*

Because the too rapid flow of cold air *reduces the temperature* of the burning mass.

263. *Why will a piece of paper twisted like an extinguisher put out a candle?*

Because, before the flame of the candle can ignite the paper, *the oxygen contained within it is consumed*, and the flame is suffocated.

"When his candle shined upon my head, and when by his light I walked through darkness."—JOB XXIX.

264. *Why do tallow candles require snuffing?*

Because the *oxygen* of the air cannot reach the wick through the body of flame—therefore the *unconsumed carbon* accumulates upon the wick.

265. *Why do composite and wax candles not require snuffing?*

Because their wicks are made by a series of plaits, by which they are bent to meet the *oxygen* of the air, and consumed.

266. *Why does setting a glass upon a lamp increase its brilliancy, though it shortens the flame?*

Because it conducts an increase of air to the flame, and the greater supply of *oxygen* causes the escaping vapour of oil to be all rapidly consumed.

267. *Why does a candle burn dimly when the wick has become loaded with carbon?*

Because the carbon *radiates* the heat, and disperses it, and reduces the heat of the flame below that temperature which is essential to its *luminosity.*

268. *What differences characterise the combustion of carbon and of hydrogen?*

The combustion of *carbon* takes place without the production of flame. The charcoal (or carbon in any other form) being heated to redness, enters directly into combination with the *oxygen* of the surrounding air, and the carbonic acid gas, being invisible, passes away unobserved.

But in the combustion of *hydrogen* the heat developed is so intense as to render *the gas itself luminous,* just as iron may be heated to a red or white heat.

269. *What has become of the candle when it has been burnt?*

It has been resolved partly into *carbonic acid gas* which, though unperceived, has diffused itself through the surrounding air; and partly into *water,* which escaped in the form of thin vapour.

270. *Has any part of the candle been consumed or lost?*

"I know that whatsoever God doeth, it shall be for ever: nothing can be put to it, nor anything taken from it; and God doeth it that men should fear before him."—ECCLES. III.

No; there is no such thing as "loss" in the operations of nature Every particle of the candle, now invisible, exists either in the form of *gas, vapour*, or *water*, with, perhaps, a few solid particles that may be called *ashes*, but which are too minute to excite attention.

The economy of nature should teach us a very impressive lesson—*nothing is suffered to be wasted*, not even the slightest atom. As soon as any body has fulfilled its purpose in one state of being, it is passed on to another. The candle, existing no longer as a candle, is flying upon the wings of the air as *carbonic acid gas*, and as *water*. These probably find their way to the garden or the field, where the carbonic acid gas forms *the food of the plant*, and the water affords it a refreshing *drink*. And can it be supposed that the Almighty Being, who has thus economised the existence of the *material* creation, should be less mindful of the immaterial *soul* of man? There *is* an eternity before us, the certainty of which is evidenced even by the laws of the material creation.

CHAPTER XIV.

271. *What is coal?*

Coal is a "*vegetable fossil.*

272. *What is meant by a vegetable fossil?*

It is a substance *originally vegetable*, which, by pressure and other agencies within the earth, has been brought to a condition approaching that of *mineral* or earthy matter.

273. *Why do we know that coal is of vegetable origin?*

By the *chemical components* of its substance; and also by the *vegetable forms* that are found abundantly in coal beds.

Professor Buckland, in his *Bridgewater Treatise*, speaking of the impressions of plants found in the coal mines, says: "The finest example I have ever witnessed is that of the coal mines of Bohemia. The most elaborate imitations of living foliage upon the painted ceilings of Italian palaces bear no comparison with the beauteous profusion of extinct vegetable forms with which the galleries of these instructive coal mines are overhung. The roof is covered as with a canopy of gorgeous tapestry, enriched with festoons of most graceful foliage, flung in wild irregular profusion over every part of its surface. The effect is heightened by the contrast of the coal-black colour of these vegetables with the light ground-work of the rock to which they are attached. The spectator feels himself transported, as if by enchantment, into the forests of another world; he beholds trees, of forms and characters now unknown upon the surface of the earth, presented to his senses almost in the beauty and vigour of their primeval life: their scaly stems and bending branches, with their delicate apparatus of foliage, are all spread forth before him, little impaired by the lapse of countless ages, and bearing faithful records of extinct systems of vegetation which began and terminated in times of which these relics are the infallible historians."

274. *What are the chemical components of coal?*

They consist of *carbon, hydrogen, oxygen*, and *nitrogen*. The proportions of these elements vary in different kinds of coal. Carbon is the chief component; and the proportions may be stated to be, generally, *carbon*, 90 per cent.; *hydrogen*, from 3 to 6 per cent.; the other elements enter into the compound in such small proportions, that, for all ordinary purposes, it is sufficient to say that coal consists of *carbon* and *hydrogen*, but chiefly of *carbon*.

275. *What is charcoal?*

Charcoal consists almost entirely of *carbon*. It is made from *wood* by the application of heat, without the admission of air. The hydrogen and oxygen of the wood are expelled, and that which remains is charcoal, or *carbon* in one of its purest states.

276. *What is animal charcoal?*

Animal charcoal, like vegetable charcoal, consists of *carbon* in a state approaching purity. It is made from the *bones of animals*, heated in iron cylinders. It is commonly called *ivory black*.

277. *What is the purest form of carbon known?*

The purest form of *carbon* is the *diamond*, which may be said to be absolutely pure.

Hence we derive another of the beautiful lessons of science—a lesson which teaches us to *despise nothing that God has given*. The soot which blackens the face of a chimney-sweep, and the diamond that glistens in the crown of the monarch, consist of the same element in merely a different atomic condition. What a lesson of humility this teaches to Pride! The haughty beauty as she walks the ball-room, inwardly proud of the radiance of her gems as they rise and fall upon her breast, little thinks or knows that *every breath that is expired around her wafts away the like element of which her treasures are composed*. That even in our own flesh and bones the same abounding substance lies hid; and that the buried tree of the primitive world, and the little flower of to-day, are both the instruments of giving this singular element to man!

278. *What is coke?*

Coke is coal, divested of its hydrogen and other volatile parts, by a similar process to that by which charcoal is produced. It forms the residue after hydrogen gas has been made from coals. It consists almost entirely of *carbon.*

279. *Why do burning coals produce yellow flame?*

Because the *hydrogen* which they contain is combined with some proportion of *carbon,* which imparts a bright yellow colour to the flames.

280. *Why do some of the flames of a fire appear much whiter than others?*

Because the quality of coals, and the conditions under which they are burnt, are liable to variation. Some coals yield a *heavy* hydrogen, called *bi-carburetted hydrogen,* which burns with a much brighter flame than *carburetted hydrogen.*

281. *Why does bi-carburetted hydrogen burn with a whiter flame than the common coal gas?*

Because it is combined with a larger proportion of *carbon,* to which it owes its increased luminosity.

282. *Why do some of the flames of a fire appear blue?*

Because the hydrogen which is escaping where those flames occur is *pure hydrogen,* destitute of carbon.

283. *Why does the fire sometimes appear red, and without flame?*

Because the volatile gases have been driven off and consumed, and combustion is continued by the *carbon* of the coals and the *oxygen* of the air.

284. *What effect has the burning of a fire upon the composition of the air?*

It is found that in burning 10lb. of coal the oxygen contained in 1,551 cubic feet of air is altogether absorbed. It is therefore necessary to keep the atmosphere of a room, in which a coal fire is burning, fresh and pure, to supply 155 cubic feet of fresh air for every pound of coal that is consumed.

285. *Why does wood which is "green" hiss and steam when it is burnt?*

Because it contains a large amount of water, which must be evaporated before combustion can proceed.

286. *What is the effect of this evaporation?*

A great deal of heat is unprofitably expended in driving off the water of the fuel.

287. *Why does poking a fire cause it to burn more brightly?*

Because it opens avenues through which the air may enter to supply *oxygen.*

288. *Why do "blowers" improve the draft of air through a fire?*

Because, by obstructing the passage of the current of air *over* the fire, they cause additional air to pass *through* it, and therefore a greater amount of *oxygen* is carried to the coals.

289. *What is smoke?*

Unconsumed particles of *coal,* rendered volatile by heat, and driven off.

290. *What is soot?*

Carbon in minute particles, driven off with other volatile matters and deposited on the walls of chimnies.

291. *Why do fresh coals increase the quantity of smoke?*

Because they contain volatile matters which are easily driven off; and because, also, they reduce momentarily the heat, so that those matters that first escape cannot be consumed.

292. *Why do charcoal and coke fires burn clearly and without flame?*

Because the *hydrogen* has been previously driven off from those substances.

293. *Why is it difficult to light charcoal and coke fires?*

"He hath made his wonderful works to be remembered: the Lord is precious and full of compassion."—PSALM CXI.

Because they contain no *hydrogen* to produce *flame*, and assist combustion.

A new plan of kindling fires has lately been recommended. Coals are to be laid in the *bottom* of the fire-place to a considerable depth, then the paper and wood are to be laid on, and then a little coals and cinders over them. This plan of "laying in" the fire is precisely the *reverse* of that which has been pursued for many years. The theory is, that when the coals in the bottom are ignited, a more even combustion is kept up, whilst the smoke and gas which would otherwise escape, and become as so much waste fuel, is burnt up, and produces heat. We have heard the plan strongly recommended by persons who have tried it, and who testify to the great economy of fuel to which it conduces.

CHAPTER XV.

294. *Why does paper ignite more readily than wood?*

Because its texture is less *dense* than that of wood; its particles are therefore more *readily heated* and decomposed.

295. *But if articles of loose texture are bad conductors of heat, why do they so easily ignite?*

The fact that they are *bad conductors* assists their ignition. The heat which would pass from particle to particle of the dense substance of iron, and be *conducted away*, accumulates in the interspaces of paper, and ignites it.

296. *Why does wood ignite less readily than paper?*

Because its substance is *denser* than that of paper; it therefore requires a higher degree of heat to inflame its substance.

297. *Why does wood, when ignited, burn longer than paper?*

Because, being a denser substance, it submits a *larger number of particles*, within a given space, to the action of the heat, and the formation of gases.

298. *Why do we, in lighting a fire, first lay in paper, then wood, and lastly coals?*

Because the paper is more easily ignited than wood, and wood than coals; therefore the *paper* assists the ignition of the *wood*, and the *wood* assists the ignition of the *coals*.

299. *Why will not wood ignite by the flame of a match?*

It will do so, unless there is a great disproportion between the size of the wood and the flame of a match. A *thin* piece of wood will ignite, but a square block will not, because the heat of the flame is insufficient to raise the temperature of a *large surface* to the point that will drive out its gases

300. *Why do we place the paper under the wood, and the wood under the coals?*

Because heat and flame, when surrounded by air, have a strong tendency to spread themselves *upwards*.

301. *Would it be possible to light the coals by putting the paper and the wood upon the top?*

It would be possible; but the loss of heat would be so great, that a *much larger quantity* of paper and wood would be required.

302. *Why does a poker laid across the top of a dull fire revive it?*

Because the poker *radiates* the heat it receives from the fire downward upon the fuel.

Because, also, it divides the ascending air, and thereby *creates currents*.

The amount of good which the poker does to the fire is very slight indeed. Generally, the housewife stirs the fire first, and blows or brushes away the ashes that prevent the influx of air. She then places the poker upon the top, and the popular mind supposes that the poker "draws" the fire. The custom of placing a poker over the fire is of very remote antiquity. It was once believed that forming *a cross*, by placing the poker over the bars, protected the fire from the hostility of malignant *witches!*

303. *Why should fire-places be fixed as low as possible in rooms?*

Because heat *ascends*, and when the fire-places are high the lower parts of the room are *inadequately warmed*.

Also, as currents of air fly towards the fire, elevated fire-places *cause drafts* about the persons of the inmates to a much greater extent than they would if they were lower down.

304. *Why, if a piece of paper be laid with its flat surface upon the fire, will it "char," but not ignite?*

Because, as in the case of the proper candle-extinguisher, the *carbonic acid gas* accumulating beneath it prevents its igniting.

305. *Why, if you direct a current of air towards the paper, will it burst into a blaze?*

Because the carbonic acid gas is displaced by a current of air containing *oxygen.*

306. *Why does water extinguish fire?*

Because it *saturates the fuel,* and prevents the gases thereof from combining with the oxygen of the air.

307. *As water contains oxygen, why does not the oxygen of the water support the fire?*

Because the affinity between the *hydrogen* and *oxygen* of the water is so strong that fire cannot separate them.

Water may be decomposed by *heat,* as will be hereafter explained. But the heat of an ordinary fire is insufficient. There is, however, some reason for believing that, in cases of very large fires, such as the accidental burning of houses, &c., when the supply of water thrown upon the fire is very deficient. the water *does* become *decomposed,* and add to the fury of the flames.

308. *Why does the blacksmith sprinkle water upon the coals of his forge?*

The blacksmith uses *small coals* because the small pieces thereof are more easily ignited than large lumps would be, and they convey heat better by completely surrounding the articles put into the fire. He sprinkles water on the coal dust *to hold its particles together by cohesion,* until the heat forms it into a cake. A strong blast of hot hair drives the vapour of the water away, and leaves a porous mass to the action of the fire.

309. *Why, when the blacksmith thrusts a heated iron into a tankard of water, do we recognise a peculiar smell?*

"Oh the depth of the riches both of the wisdom and knowledge of God! how unsearchable are his judgments, and his ways past finding out."—ROM. XI.

Because the intense heat disengages a small volume of the gases of which water is formed.

310. *Which gas do we (in this instance) recognise by the smell?*

The *hydrogen* gas. Oxygen gas possesses no odour.

311. *What is Spontaneous Combustion?*

Spontaneous combustion is that which occurs in various bodies when they become highly heated by *chemical changes.*

312. *Why is heat developed during chemical changes?*

Because, as all bodies contain *latent caloric,* the disturbance of the atoms of which those bodies are composed, during the new combinations that constitute *chemical changes,* frequently sets the caloric free, and an *accumulation of caloric* produces spontaneous combustion.

313. *Does a match ignite spontaneously when drawn over a rough surface?*

No. Because in this case the combustion arises from heat *applied by friction.*

314. *Does phosphorous ignite spontaneously when held in a warm hand?*

Phosphorous will ignite when held in a warm hand, but it does not then produce spontaneous combustion, because it ignites through the agency of *applied heat.*

315. *But if a piece of dry phosphorous be sprinkled with powdered charcoal it will ignite, without the application of heat. Why is this?*

Because the *carbon* (charcoal) absorbs *oxygen* from the air, and conveys it to the *phosphorous.* Here are *chemical changes* which develope heat, and produce *spontaneous combustion.*

316. *Why do hay stacks sometimes take fire?*

Because the hay, having become damp, decays, and passes on to a state of *fermentation,* in which *chemical changes occur,* during

"Who hath woe? who hath sorrow? who hath contentions? who hath babbling? who hath words without cause? who hath redness of the eyes? * * * They that tarry long at the wine."—PROV. XXIII.

which heat is evolved. Hay, taking fire under these circumstances, would exhibit *spontaneous combustion.*

317. *What substances are liable to produce spontaneous combustion?*

All substances which contain sugar, starch, and other components liable to *fermentation.* All bodies that evolve, under low degrees of temperature, *inflammable gases.* And all organic bodies undergoing decay.

Grain, cotton, hemp, flax, coals, oily and greasy substances.

318. *What is the Ignis Fatuus (sometimes called "Will-o'-the-Wisp," "Corpse Candles," and "Jack-o'-Lantern")?*

It is a flame produced by spontaneous combustion, caused by the decay of animal or vegetable bodies, which evolve *phosphoretted hydrogen* gas, under circumstances attended by a low degree of heat, sufficient to ignite the gases. It is mostly seen over marshy places, and burial-grounds.

Many a "Ghost Story" has owed its origin to these singular but harmless appearances. People, ignorant of the cause, have been terrified at the effect. To the fancy of an affrighted mortal, the simple flame of the *Ignis Fatuus* has assumed the form of a departed friend, and even found a supernatural voice. If, excited by a momentary daring, the beholder moved towards the light upon which he gazed, it fled from him. If he turned from it and walked away, it followed him, step by step. The darkness of a lonely road, or the sacred solitude of a burial-place, have been sufficient accessories to authenticate the appearance of a spirit. And yet how simple the phenomenon? Matters so volatile as those which produce the *Ignis Fatuus* would naturally be driven back by the motion in the air caused by an advancing body; and, on the other hand, a body moving from them would create a current in which the *Ignis Fatuus* would follow. Poisonous gases, escaping from decaying bodies, pass into the air and take fire. They are thereby converted into harmless compounds. Thus we see that the "ghost" which terrifies the mind of the ignorant, becomes a "guardian angel" to the educated.

319. *Has spontaneous combustion ever occurred in living bodies?*

It has occurred in numerous instances to persons habituated to the excessive use of spirits.

320. *Why should spontaneous combustion occur in the case of the drunkard?*

"Drought and heat consume the snow waters; so doth the grave those which have sinned."—JOB XXIV.

Because spirituous drinks contain a large proportion of *alcohol*, one of the constituents of which is *hydrogen.* The vital energies of the drunkard, being destroyed by excess, chemical agencies obtain an ascendancy, and it is supposed that the *hydrogen* of the alcohol combines with the *phosphorous* of the body to form *phosphoretted hydrogen*, which ignites spontaneously, and literally consumes the living temple.

Cases of spontaneous combustion are of rare occurrence. But they are sufficiently well authenticated by high medical authority, in many parts of the world, to present an awful warning to the inveterate drunkard. The cases of which we have read the particulars present details of the most appalling description. How signally the Almighty displeasure at intemperance is expressed, when the very drink which imparts the mad pleasure of intoxication is made the *direct* instrument by which the drunkard is destroyed!

CHAPTER XVI.

321. *Why does friction produce heat?*

Because all bodies contain *latent heat*, that is, heat that lies hid in their substance, and the rubbings of two bodies against each other *draws the latent heat to the excited surfaces.*

322. *Why does the rubbing of two surfaces together attract latent heat to those surfaces?*

Because it is a law of nature that *heat* shall always attend *motion;* and it is generally found that the *intensity of heat* bears a specific relation to the *velocity of motion.*

323. *What are the sources of heat?*

The *rays* of the *sun*, the *currents* of *electricity*, the *action* of *chemicals*, and the *motion* of *substances.*

324. *Why does water freeze?*

Because its latent heat is partly *drawn off* by the surrounding air.

325. *Why does ice melt?*

Because the heat, once latent in the water, but drawn off by the air, *has returned* to it, and restored the water to its former condition.

326. *Why does water become steam?*

Because a larger amount of heat has entered into it than can remain latent in water. The water therefore expands and rises in the form of vapour, or *water attenuated by heat.*

327. *How many degrees of heat are latent, or hidden, in the different states of water?*

In thawing *ice,* 140 deg. of caloric become latent; and in converting the water into steam, 1,000 deg. more of caloric are be taken up. Therefore, *ice* requires to take up 1,140 deg. of latent caloric before it becomes steam.

328. *What is the most modern theory of heat?*

It is this—that caloric, which produces heat, is an extremely *subtile fluid,* of so refined a nature that it possesses no weight, yet is capable of diffusing itself among the particles of the most solid bodies.

It is also believed that—all bodies are subject to the action of two opposing forces: one, the *mutual attraction* of their *particles;* the other, the *repulsive force* of *caloric*—and that bodies exist in the *æriform, fluid,* or *solid state, according to the predominance of either the one or the other of these opposing forces.*

329. *How do we measure the quantity of caloric in any substance?*

It is impossible to determine the amount of caloric which any body contains. Our *sensations* would obviously be deceptive, since, if we dipped the right hand in snow, and held the left hand before the fire, and then immersed both hands in cold water the water would feel *warm* to the *right hand* and *cold* to the *left hand.*

But, as *caloric* uniformly expands substances that are under its influence, one of the bodies most sensitive to *calorific* effects has been selected to be the *indicator* of the amount of *caloric.* This substance is *quicksilver;* and the scale of measurement, and the apparatus for exhibiting the rise or fall of the quicksilver, constitute the *thermometer.*

330. *If it is impossible to measure the amount of caloric in any substance, how can it be said that ice absorbs 140 deg. in becoming water?*

Those figures simply record the amount of caloric indicated by the *thermometer*. The instrument will show with sufficient accuracy the *relative amount* of caloric in various bodies, or in the same bodies *under different circumstances*, but it can never determine the *precise amount of caloric* in any one body.

331. *Why, if a hot and a cold body were placed near to each other, would the cold one become warmer, and the hot one cooler?*

Because *free caloric* (that is, caloric that is not latent,) always exhibits a tendency to establish an *equilibrium*. If twenty bodies, of different temperatures, were placed in the same atmosphere, they would *all soon arrive at the same temperature*. The caloric would leave the bodies of those of the *highest*, and find its way to those of the *lowest* temperature.

332. *How does caloric travel?*

It travels in *parallel rays* in all directions with a velocity approximating to that of light; and it passes through various bodies with a rapidity proportionate to their power of *conduction*.

333. *Why does melted metal run like a stream of fluid?*

Because *caloric* has passed into its substance, and, repelling its particles, has separated them to that degree which produces fluidity.

334. *How do we know that it is caloric passing into the substance of the metal which produces this effect?*

Because, as soon as a bar of metal begins to be heated, it *expands* and *lengthens*. It continues to do so, until the heat arrives at that point which *causes the metal to melt*.

335. *Why does the iron of an ironing-box sometimes become too large for the box to receive it?*

Because *caloric* has passed into the substance of the iron, and *repelled its particles*, by which it has become expanded.

336. *Why does the iron enter the box when it has become partially cooled?*

Because a portion of the caloric has left the iron, the particles of which have *drawn closer together*, and contracted the mass.

This effect is frequently observed by females in domestic life, who, when they are ironing, or using the Italian irons, find that the heated metal has been too much expanded to enter the box or tube. They find it necessary to wait until the cooling of the iron has had the effect of reducing its dimensions. The expansion of bodies by heat is one of the grandest and most important laws of nature. We are indebted to it for some of the most beautiful, as well as the most awful, phenomena. And science has gained some of its mightiest conquests through its aid. Yet frequently, though quite unthought of, in the hands of the humble laundress, will be found a most striking illustration of this wonderful force of caloric.

337. *Are there any instances in which the abstraction of latent heat will reduce the bulk of bodies?*

Yes, there are several. But the most familiar one is that which is exhibited by mixing a *pint* of the *oil of vitriol* with a *pint* of *water. A considerable amount of heat will be evolved;* and it will be found that the two pints of fluid *will not afterwards fill a quart measure.*

338. *Is there any latent heat in air?*

Yes: a considerable amount. In a pint measure of air, though in no way evident to our perceptions, there lurks sufficient caloric to raise a piece of metal several inches square to glowing redness.

339. *How do we know that caloric exists in the air?*

It has been positively demonstrated by the invention of a small condensing syringe, by which, through the rapid compression of a small volume of air, a spark is emitted which ignites a piece of prepared tinder.

340. *What is the cause of the spark when a horse's shoe strikes against a stone?*

The *latent heat* of the iron or the stone is set free by the *violent percussion.* The same effect takes place when *flint* strikes against *steel*, as in the old method of obtaining a light with the aid of the tinder-box.

What an eloquent lecture might be delivered upon the old-fashioned tinder-box, illustrated by the one experiment of "striking a light." In that box lie, cold and motionless, the Flint and Steel, rude in form and crude in substance. And yet, within the breast of each there lies a spark of that grand elem[illegible] which influences every atom of the universe; a spark which could invoke the fierce agents of destruction to wrap their blasting flames around a stately forest, or a crowded city, and sweep it from the face of the world; or which might kindle the genial blaze upon the homely hearth, and shed a radiant glow upon a group of smiling faces; a spark such as that which rises with the curling smoke from the village blackmith's forge—or that which leaps with terrific wrath from the troubled breast of a Vesuvius. And then the tinder—the cotton—the carbon: What a tale might be told of the cotton-field where it grew, of the black slave who plucked it, of the white toiler who spun it into a garment, and of the village beauty who wore it—until, faded and despised, it was cast amongst a heap of old rags, and finally found its way to the tinder-box. Then the Tinder might tell of its hopes; how, though now a blackened mass, soiling everything that touched it, it would soon be wedded to one of the great ministers of nature, and fly away on transparent wings, until, resting upon some Alpine tree, it would make its home among the green leaves, and for a while live in freshness and beauty, looking down upon the peaceful vale. Then the Steel might tell its story, how for centuries it lay in the deep caverns of the earth, until man, with his unquiet spirit, dug down to the dark depths and dragged it forth, saying, "No longer be at peace." Then would come tales of the fiery furnace, what Fire had done for Steel, and what Steel had done for Fire. And then the Flint might tell of the time when the weather-bound mariners, lighting their fires upon the Syrian shore, melted silicious stones into gems of glass, and thus led the way to the discovery of the transparent pane that gives a crystal inlet to the light of our homes; of the mirror in whose face the lady contemplates her charms; of the microscope and the telescope by which the invisible are brought to sight, and the distant drawn near; of the prism by which Newton analysed the rays of light; and of the photographic camera in which the sun prints with his own rays the pictures of his own adorning. And then both Flint and Steel might relate their adventures in the battle-field, whither they had gone together; and of fights they had seen in which man struck down his fellow-man, and like a fiend had revelled in his brother's blood. Thus, even from the cold hearts of flint and steel, man might learn a lesson which should make him blush at the "glory of war;" and the proud, who despise the teachings of small things, might learn to appreciate the truths that are linked to the story of a "tinder-box."

LESSON XVII.

341. *Since all bodies expand by heat and contract by cold, why does water, when it reaches the freezing point, expand?*

Because, in freezing, water undergoes crystallization, in which its particles assume a new arrangement occupying *greater space.*

342. *Why does water never freeze to a great depth?*

Because the covering of ice which is formed upon the surface of

"For he saith to the snow, Be thou on the earth; likewise to the small rain, and to the great rain of his strength."—JOB XXXVII.

the water prevents the cold air from continuing to draw off the *caloric* of the water.

343. *Why has this exceptional law of the expansion of water, when freezing, been ordained?*

Because, but for this, deep waters might be frozen through their whole depth. This would destroy the myriads of fish and other living things that inhabit the water. Parts of the earth, now clad in verdure, would be lost in eternal winter; and even in the most temperate zones it would take months to effect a thaw; and thawing would be attended with such floods and subterranean commotions as are terrible to contemplate.

344. *Why are bed-room windows sometimes covered with crystalline forms on winter mornings?*

Because the vapour of the breaths of the inmates has condensed upon the window-panes, and formed water. The water has frozen with the cold, and exhibits the beautiful crystalline forms into which its particles are arranged.

[Here we have another domestic illustration of the great laws of nature. It is the same law which locks the artic regions in ice and decorates our window-panes. This beautiful phenomenon is usually witnessed by us on frosty mornings when we rise from our beds. It has a story which the observer of nature may read in its sparkling eyes. It tells that, although without the air is biting cold, God has wrapped a mantle around the face of nature to keep it from injury; and that the earth and the waters, though looking chilled and dead, have still the warmth of life preserved in their bosoms.]

345. *What is dew?*

Dew is *watery vapour* diffused in the air, *condensed* by coming in contact with bodies *colder than the atmosphere.*

346. *Why does the air become charged with watery vapour?*

Because, during the day, under the influence of the sun's rays, vapours are *exhaled* from all the moist and watery surfaces of the earth. These vapours are *held in suspension* in the atmosphere until, by a change in the temperature of the earth, and of bodies on the surface of the earth, they are *condensed,* and deposited in translucid drops.

347. *What causes the decline of temperature that favours the deposition of dew?*

The earth, which during the day *received heat* from the solar rays, *radiates the heat* back into the air, and therefore becomes itself colder. All the various objects upon the face of the earth also *radiate heat* in a greater or lesser degree. And dew will be found to be deposited upon the surfaces of such bodies in proportion to the fall of their temperature through *radiation.*

348. *Why is there little or no dew when the nights are cloudy?*

Because clouds act as secondary radiators; and when the *earth* radiates its heat towards the *clouds,* the clouds again *radiate it back to the earth.*

Fig. 3.—ILLUSTRATING THE FORMATION OF DEW.

If plates of glass be laid over grass-beds, as in the engraving Fig. 3, no dew will be deposited on the grass underneath the glass plates, although all around the grass will be completely wetted. The explanation is that the glasses, being radiators of heat, act in the same manner as the clouds, returning the heat to the bodies underneath them, and preventing the formation of dew thereon.

349. *Why does dew form most abundantly on cloudless nights?*

Because the heat which is radiated by the earth does not return to it. The temperature of the earth, and the air immediately upon its surface, is therefore lowered, and dew is formed.

It has been observed that sheep that have lain on the grass during the formation of dew have their backs completely saturated with it, but that underneath the line where their bodies turn to the earth, their coats will be dry. In the same manner glass globes suspended in the air, on dew forming nights, will be found loaded with globules of dew upon the top, but there will be no appearance of moisture underneath.

350. *Why are star-lit nights usually colder than cloudy nights?*

Because heat is *radiated* from the earth, and passes away into the utmost regions of the atmosphere.

351. *Why is there little dew under branches of thick foliage?*

Because the foliage *acts as a screen*, which prevents the radiated heat of the earth from passing away.

352. *Why is there no dew formed on windy nights?*

Because, as winds generally consist of dry air, they *absorb and bear away* the atmospheric moisture.

353. *Why are valleys and low places chiefly subject to dew?*

Because the elevated lands around them *prevent the disturbance of the air* in which the moisture is held.

354. *What bodies are most likely to be covered with dew?*

All bodies that are *good radiators of heat*, such as wool, swans-down, grass, leaves of plants, wood, &c.

355. *What bodies are likely to receive little dew?*

All *bad radiators of heat*, such as polished metal surfaces, smooth stones, and polished surfaces generally. Dew will be found to lie more abundantly upon rough and woolly leaves than upon smooth ones.

356. *At what period of the night is the largest amount of dew usually formed?*

It is generally supposed that dew is formed *most copiously* in the *mornings* and *evenings*. But *such is not the case*. It is deposited at all hours of the night, but *most plentifully after midnight*.

357. *Why is dew formed most plentifully after midnight?*

Because, as *radiation* has been going on for some time, the temperature of the earth, and of various bodies upon it, has been *considerably reduced*.

"Out of whose womb came the ice? and the hoary frost of heaven, who hath gendered it?"—JOB XXXVIII.

358. *In what parts of the world is the maximum of dew formed?*

In warm lands near the sea, or in the vicinity of rivers or lakes, as the localities of the Red Sea, the Persian Gulf the coast of Coromandel, in Alexandria, and Chili.

359. *In what parts of the world is the minimum of dew formed?*

It is quite absent in arid regions, in the interior of continents, such as Central Brazil, the Sahara, and Nubia.

360. *Why is dew seldom formed at sea?*

Because of the defective *radiating* quality of the surface of *water.*

361. *Why is a heavy dew regarded as the precursor of rain?*

Because a heavy formation of dew indicates that the air is *saturated with moisture.*

362. *What is hoar-frost?*

Hoar-frost is frozen dew.

363. *Why is hoar-frost said to foretell rain?*

Because it shows that the air is saturated with moisture, and the temperature of the air being low, the vapours are *likely to condense,* and produce *showers.*

364. *What is honey-dew?*

Honey-dew is the name applied to a *sweet and sticky moisture* occasionally deposited upon the leaves of plants. It is, however, an error to call it *dew,* as it is procured by a class of *insects* termed *aphides.*

365. *What are fogs?*

Fogs are *clouds* formed near the earth's surface; but London fogs are distinguished from clouds by the fact that they embrace in their vaporous folds the *smoke* and *volatile matters* imparted to the air by the operations of man. This is also the case with fogs generally that arise near large towns.

"Hath the rain a father? or who hath begotten the drops of dew?"—
JOB XXXVIII.

366. *Why are certain coasts liable to almost perpetual fogs?*

Because of local or geographical agencies which contribute to their production. The coasts of California are almost constantly wrapped in fog; and, almost as constantly, the western coast of the American continent, as far south as Peru. Newfoundland, Nova Scotia, and Hudson's Bay, are all subject to dense and frequent fogs arising from the condensation of vapour from the water flowing from the hot Gulf-stream, coming in contact with the colder air.

367. *What are dry fogs?*

Dry fogs are characterised by a dull opaque appearance of the atmosphere. They are most common in certain parts of North America, though they sometimes occur in Germany and in England. They are generally referred to the *electrical state of the atmosphere*, but the theory of them is still a matter of doubt.

368. *What is a mist?*

The term *mist* is generally applied to vapours that rise over *marshy places*, or the surfaces of *water*, and roll or move over the land.

369. *What is the difference between a mist and a fog?*

Fogs, as they are known to us, generally arise over the *land*, and are usually mingled with the smoke of large towns. Mists generally arise over water, or wet surfaces.

370. *Why do mists and fogs disappear at sunrise?*

Because the condensed vapours are again *expanded* and *dispersed* by the heat of the sun's rays.

371. *Why do fogs frequently rise in the morning and fall again in the evening.*

Because, warmed by the sun's rays, they become more rarefied, and fly away at an altitude where they *appear* to be altogether dispelled; but at night, when the earth *cools by radiation*, the vapours near the earth *again condense*, and settle in the *form of fog*.

372. *Why do fogs sometimes rest upon a given locality for several days together, and then disappear?*

"He bindeth up the waters in his thick clouds; and the cloud is not rent under them."—JOB XXVI.

They are probably kept near to the surface of the earth by a superstratum of cold air. A cold air lying *above*, or a cold air lying ***below***, might equally contribute to keep a fog near the surface of a particular part of the earth, until a *flow of wind*, or a *fall of rain*, altered the atmospheric condition.

There are many interesting facts connected with the history of dew. It has attracted the attention of natural philosophers in all ages. But its true theory was never understood until recently. The ancients imagined that dews were shed from the stars; and the alchemists and physicians of the middle ages believed that the dew distilled by night possessed penetrating and wonder-working powers. The ladies of those times sought to preserve their beauty by washing in dew, which they regarded as a "celestial wash." They collected it by placing upon the grass heaps of wool, upon the threads of which the magic drops clustered.

CHAPTER XVIII.

373. *What are clouds?*

Clouds are volumes of *vapour*, usually elevated to a considerable height.

Fig. 4.—CIRRO-CUMULUS, OR SONDER CLOUD.

374. *Whence do clouds arise?*

From the *evaporation of water* at the earth's surface.

375 *Why do we not see them ascend?*

We do, sometimes, in the form of what we call ***mists***, but generally the vapours that rise and contribute to the formation of clouds are so thin that they are *invisible*.

"With clouds he covereth the light, and commandeth it not to shine by the cloud that cometh betwixt."—JOB XXXVI.

376. *Why, if they are invisible when they rise, do they become visible when they have ascended?*

Because the vapours become *cooled* in passing through the air, and form a denser body.

377. *Why, when they are condensed, do they not follow the course of gravitation, and descend?*

Because the vapours form into *minute vesicles*, which we may call *vapour bubbles*, and these, being warmed by the sun, are specifically *lighter than the air*.

Because, also, the lower parts of clouds *do partially* descend, but again becoming more *rarefied* by meeting with a *warmer atmosphere*, they again ascend, and are thus *poised* upon the air.

Because, also, there is always a degree of atmospheric motion *upward*, caused by the *convection of heat* from the earth's surface. And, although there must also be downward movements of the air to supply the place of that which has ascended, still *the heat* of the ascending air, *combined with its upward movement*, expands and floats the vapour of the clouds.

378. *At what height do clouds usually fly?*

They fly at every degree of altitude; but clouds of *specific character* are said to fly at given altitudes, or to occupy certain *ranges of altitude*. We will give their probable altitudes when speaking of the specific clouds.

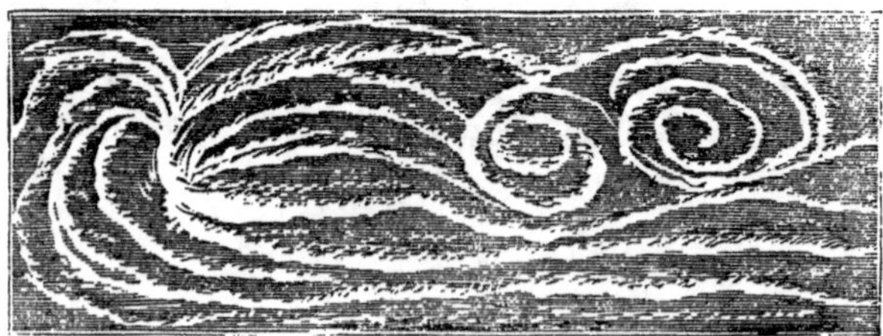

Fig. 5.—CIRRUS, OR CURL CLOUD.

379. *How many descriptions of clouds are there?*

There are *seven*.

"Who giveth rain upon the earth, and sendeth waters upon the fields."—
JOB V.

1. The *Cirrus* (Fig. 5), estimated range of altitude from 10,000 to 24,000 feet.

2. The *Cumulus* (Fig. 7), from 3,000 to 10,000 feet.

3. The *Stratus*, an extended continuous level sheet of cloud, increasing from beneath. They fly very low.

4. The *Nimbus* (Fig. 10), 1,500 to 5,000 feet.

5. The *Cirro-cumulus* (Fig. 4), from 3,000 to 20,000 feet.

6. The *Cirro-stratus* (Fig. 6), from 5,000 to 10,000 feet.

7. The *Cumulo-stratus* (Fig. 9), from 3,000 to 10,000 feet.

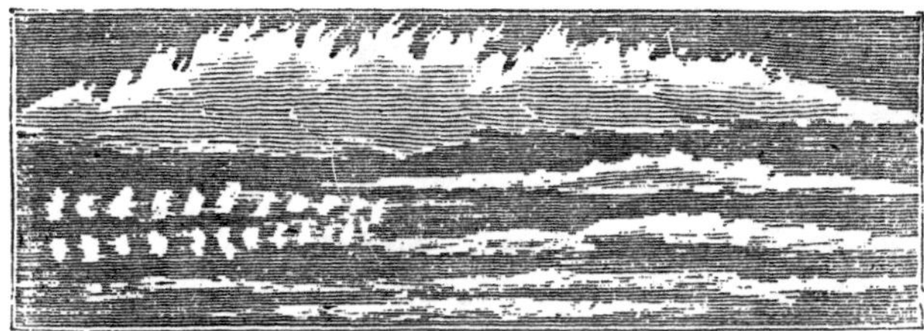

Fig. 6.—CIRRO-STRATUS, OR WANE CLOUD.

The estimated heights given must be looked upon as very conjectural although they have been derived from the best existing authorities. It is sufficient to know that the range of the altitude of the various clouds is from that of the *Nimbus*, or *thunder cloud*, 1,500 feet, to that of the *Cirrus*, 24,000 feet, the others being intermediate. The first three of the clouds above enumerated constitute what are called the *primary forms*. The remaining four are called *secondary forms*, because they arise, as their names generally indicate, out of combinations of the *primary forms*. Although, from the frequent mingling of clouds, it is not always practicable to identify them by the adopted classification, still, as there is generally a prevalence of one type of cloud over another, the observer would be able to distinguish a "*Cirrus sky*," or *Cirro-cumulus sky*," &c. Upon some occasions the typical characters of the clouds are beautifully defined; and the contemplation of their forms, and the laws of their formation, affords infinite pleasure to the observer. The advantages of scientific knowledge are such, that whether you look downwards to the earth, or upwards to the sky you have still the writing of God to read.

380. *What produces the various shapes of clouds?*

1. The state of the *atmosphere*.

2. The *electrical* condition of the clouds.

3. The *movements* of the atmosphere.

4. The *season of the year*.

"Behold, he withholdeth the waters, and they dry up; also he sendeth them out, and they overturn the earth."—JOB XII.

381. *What are the dimensions of clouds?*

A single cloud has been estimated to have as many as *twenty square miles of surface*, and to be *above a mile in thickness*, while others are no larger than a *house*, or a man's *hand*.

Fig. 7.—CUMULUS, OR PILE CLOUD.

382. *How are clouds affected by winds?*

If *cold winds* blow upon the clouds, the cold condenses the vapour, turning the clouds into *rain*. But if *warm dry winds* blow upon the clouds, they *rarefy the vapour* to a greater degree, and temporarily *disperse the clouds*.

383. *How do winds affect the shapes of clouds?*

When winds are *mild and gentle*, the clouds break into *small patches*, and rise to a considerable height. But when the winds are cold and blustering, the clouds fly low, and roll along in *heavy masses*.

384. *Why are east winds usually dry?*

Because in coming towards England they pass over vast continents of land, and comparatively little ocean. Hence they are not loaded with *vapours*.

385. *Why do west winds generally bring rain?*

Because they come across the *Atlantic*, and are heavily charged with *vapour*.

386. *Why are north winds generally cold and dry?*

Because they come from the arctic ocean, over vast areas of *ice and snow*.

"Terrors are turned upon me: they pursue my soul as the wind; and my welfare passeth away as a cloud."—JOB XXX.

387. *Why are south winds warm and rainy?*

Because they come from the southern regions, heated by the *hot earth and sands*, and as they cross the sea they *absorb a large amount of vapour.*

Fig. 9.—CUMULO-STRATUS, OR TWAIN CLOUD.

388. *Why are clouds said to indicate the changes of the weather?*

Because, as it is the *state of the clouds* that, to a great extent, determines the *state of the weather*, the formation of the clouds must predicate approaching changes.

389. *What do cirrus clouds foretell?*

Cirrus clouds foretell *fine* weather, when they fly high, and are thin and light.

They foretell *light rain* when, after a long continuance of fine weather, they form fleecy lines stretched across the sky.

They foretell a *gale of wind* when, for some successive days, they gather in the same quarter of the heavens, as if denoting the point from which to expect the coming gale. (Fig. 5).

390. *What do cumulus clouds foretell?*

Cumulus clouds, when they are well defined, and advance with the wind, foretell *fine weather.*

"When he made a decree for the rain, and a way for the lightning and the thunder."—JOB XXVIII.

When they are thin and dull, and float against the wind, or in opposition to the lower currents, they *foretell rain.*

When they increase in size, and become *dull and grey at sunset*, they predict a *thunder-storm.* (Fig. 7.)

391. *What do stratus clouds foretell?*

Stratus clouds foretell *damp and cheerless weather.*

392. *What do nimbus clouds foretell?*

Nimbus clouds foretell *rain, storm,* and *thunder.* (Fig. 10.)

393. *What do cirro-cumulus clouds foretell?*

Cirro-cumulus clouds, in summer, foretell *increasing heat* attended by *mild rain,* and a *south wind;* but in winter they commonly precede the *breaking up of a frost,* and the setting in of *foggy and wet weather.* (Fig. 4.)

394. *What do cirro-stratus clouds foretell?*

Cirro-stratus clouds foretell *rain* or *snow,* according to the season of the year.

These clouds extend in long horizontal streaks, thinning away at their base, and in parts becoming wavy or patchy.

When they are thus defined in the heavens they are a certain indication of *bad weather.* (Fig. 6.)

395. *What do cumulo-stratus clouds foretell?*

Cumulo-stratus clouds usually foretell a *change of weather*—from rain to fine, or from fine to rain. (Fig. 9.)

Fig. 10.—NIMBUS, OR STORM CLOUD.

"Behold, I will put a fleece of wool in the floor; and if the dew be on the fleece only, and it be dry upon all the earth beside, then shall I know that thou wilt save Israel. * * *

CHAPTER XIX.

396. *Why are cloudy days colder than sunny days?*

Because the clouds intercept the *solar rays* in their course towards the earth.

397. *Why are cloudy nights warmer than cloudless nights?*

Because the clouds *radiate back to the earth* the heat which the earth evolves?

Because, also, the clouds radiate to the earth the heat they have *derived from the solar* rays during a cloudy day.

398. *Why is the earth warmer than the air during sunshine?*

Because the earth freely *absorbs the heat of the solar rays;* but the air derives *comparatively little heat* from the same source.

399. *Why does the earth become colder than the air after sunset?*

Because the earth *parts with its heat freely by radiation;* but the air does not.

400. *Why do glasses, mats, or screens, prevent the frost from killing plants?*

Because they prevent the *radiation of heat from the plants,* and also from the earth beneath them.

401. *Why are the screens frequently covered with dew on their exposed sides?*

Because they radiate heat from *both their surfaces.* A piece of glass, laid horizontally over the earth, would radiate heat both *upwards* and *downwards.* But on its lower surface it would *receive* the radiated heat of the earth, while from its upper surface it would *throw off its own heat* and become cool. Therefore dew would be deposited upon the *upper*, but not on the *under* surface.

402. *Why does dew rest upon the upper surfaces of leaves?*

"And it was so: for he rose up early on the morrow, and thrust the fleece together, and wringed the dew out of the fleece, a bowl full of water.

Because the under surfaces receive the *radiated warmth of the earth.*

403. *Why are cultivated lands subject to heavier dews than those that are uncultivated?*

Because cultivation breaks up the hard surface of the earth, and thus *its radiating power is increased.*

404. *Why is the gravel walk through a lawn comparatively dry while the grass of the lawn is wet with dew?*

Because gravel is a *bad radiator*, but grass is a *good radiator.*

405. *What benefit results from this arrangement?*

In cultivated lands, where moisture is required, it is *induced* by the very necessity which demands it; while in rocky and barren places, where it would be of no good, dew *does not form.*

406. *Why does little dew form at the base of hedges and walls, and around the trunks of trees?*

Because those bodies in some degree *counteract the radiation* of heat from the earth; and they also *radiate heat* from their own substances.

407. *Why do heavy morning dews and mists usually come together?*

Because they both have their origin in the *humidity of the atmosphere.* The temperature of the earth having fallen, dew has been deposited; but, at the same time, the condensation of the vapour in the air *has formed a screen over the surface of the earth,* which has checked *the further radiation of heat,* and, consequently, *the further formation of dew.* The sun rises, therefore, upon an atmosphere charged with visible vapour at the earth's surface, and his first sloping rays, *having little power to warm the atmosphere,* the mist *continues visible for some time.*

408. *What effect have winds upon the formation of dew?*

Winds, generally, and especially when rapid, prevent the formation of dew. But those winds that are moist, and *contribute to the formation of clouds,* indirectly aid the formation of dew *through the*

"And Gideon said unto God, * * * Let it now be dry only upon the fleece, and upon all the ground let there be dew.

formation of clouds, and also by the *moisture they impart to the air.*

409. *Why does the humidity of the atmosphere sometimes form clouds, and at others form fogs, mists, dews, &c.?*

The result depends upon the varying *temperature, motion,* and *direction* of the atmosphere.

A *warm light atmosphere,* of a few day's duration, will elevate the vapours to the region where they are formed into *clouds.*

A *chill air,* lying upon the surface of the warmer earth, will occasion *mists* or *fogs.*

A *cold earth,* acting upon the vapours contained in a *warmer atmosphere,* will condense them and occasion *dews.*

410. *Why are frosty mornings usually clear?*

Because, in the cold atmosphere which preceded the frost, *there was but little evaporation;* and now that the frost has set in, the vapours that existed have become *frozen* in the form of *hoar-frost.*

411. *Why are clear nights usually cold?*

Because the "screen" afforded by the clouds does not exist; therefore the heat of the earth escapes, while the vapours of the air are abstracted from it by condensation into dew, thereby imparting great *clearness to the nights.*

412. *Why are hoar-frosts, or, as they are termed, "white frosts," so frequent, and "black frosts" so unusual?*

Because white, or *hoar-frosts,* result from the *coldness of the earth,* which, from its great radiating power, is always varying. But *black-frosts* result from the *coldness of the air,* which is liable to less variation of temperature than the earth.

413. *What is a black-frost?*

A *black-frost* results from the *coldness of the atmosphere,* which is at the time overshadowed by a dull cloud, giving a darkness to everything, and a leaden appearance to the *frozen surface of water.*

414. *Why are black-frosts said to last?*

"And God did so that night: for it was dry upon the fleece only, and there was dew on all the ground."—JUDGES VI.

Because as they result from the temperature of the air, which is less likely to vary than that of the earth, there is a probability that the coldness thereof *will last for some time.*

415. *What benefits result from the radiation of heat, &c.?*

But for the *radiation of heat*, we should be subjected to the most unequal temperatures. The setting of the sun would be like *the going-out of a mighty fire.* The earth would become *suddenly cold*, and its inhabitants would have to bury themselves in warm covering, to wait the return of day. By the *radiation* of heat, an *equilibrium of temperature* is provided for, without which we should require a new order of existence.

The amount of heat which our earth *receives from the sun*, and the economy of that heat by the laws of *radiation, reflection, absorption*, and *convection*, are exactly proportionate to the necessities of our planet, and the living things that inhabit it. It is held by philosophers that any change in the orbit of our earth, which would either increase or decrease the amount of heat falling upon it, would, of necessity, be followed by the *annihilation of all the existing races.* The planets Mercury and Venus, which are distant respectively 37 millions of miles, and 68 millions of miles, from the great source of solar heat, possess a temperature which would *melt our solid rocks;* while Uranus (1,800 millions of miles), and Neptune (whose distance from the sun has not been determined), must receive so small an amount of heat, that water, such as ours, would become as solid as the hardest rock, and our atmosphere would be resolved into a liquid! Yet, poised in the mysterious balance of opposing forces, our orb flies unerringly on its course, at the rate of 65,000 miles an hour; preserving, in its wonderful flight, that precise relation to the sun, which takes from his life-inspiring rays the exact degree of heat, which, being shared by every atom of matter, and every form of organic existence, *is just the amount needed to constitute the heat-life of the world!*

CHAPTER XX

416. *What is rain?*

Rain is the *vapour of the clouds* which, being condensed by a fall of temperature, forms drops of water that descend to the earth.

It is the *return to the earth* in the form of *water*, of the moisture *absorbed by the air* in the form of *vapour.*

417. *Does rain ever occur without clouds?*

It sometimes, but rarely happens, that a sudden transition from warmth to cold will *precipitate the moisture of the air,* without the formation of *visible clouds*

418. *Why are drops of rain sometimes large and at other times small?*

Because the drops, in falling, *meet and unite,* and also gather *moisture* in their descent. The greater the height from which a rain drop has descended, *the larger it is,* provided that its whole course lay through a *rainy atmosphere.*

The size of the drops is also influenced by the *amount of moisture in the atmosphere,* the *degree of cold,* and the *rapidity* of the *change of temperature,* by which the drops are produced.

419. *In what seasons of the year are rains most prevalent?*

Throughout *Central Europe* rains are most prevalent in *summer,* but in *Southern Europe* the preponderance is on the side of *winter rains.*

420. *In what months of the year does it rain most frequently in this country?*

It rains more frequently *from September to March,* than from *March to September;* but the *heaviest rains* occur from *March to September.*

421. *Why are there more rainy days from September to March?*

Because the temperature of the air is more frequently lowered to that degree which *precipitates its vapours.*

Months in the order of their comparative wetness:—1. October. 2. February. 3. July. 4. September. 5. January. 6. December.

Months in the order of their comparative dryness:—1. March. 2. January 3. May. 4. August. 5. April. 6. November.

422. *In what part of the world does the greatest quantity of rain fall?*

The greatest *quantity* of rain falls near the *equator,* and the amount *decreases towards the poles.*

423. *In what part of the world do the heaviest rains occur?*

The *heaviest* rains occur in the *tropics*, during the hot season. The drops of rain in the tropical regions are so large, and the force with which they descend so great, that their splash upon the skin causes a *smarting sensation.*

424. *In what parts of the world do the least rains occur?*

There are some parts of the earth which are *rainless*, such as Egypt, the desert of Sahara, the table lands of Persia and Montgolia, the rocky flat of Arabia Petræ, &c.

425. *How many rainy days are there in a year?*

The frequency of rainy days is greatest in countries near the sea, and their number decreases the further we journey from the sea-border towards the inland. In England it rains on an average 152 to 155 days in the year.

426. *In what part of England does the greatest amount of rain fall?*

In the town of *Keswick*, in Cumberland, where 63 inches of rain fall in a year; Kendal, in Westmoreland, 58 inches; Liverpool, 34 inches; Dublin, 25 inches; Lincoln, 24 inches; London, 21 inches.

427. *Why do the heaviest rains occur at the tropics?*

Because the *hot air* absorbs a large amount of vapour, and rises into the higher regions of the atmosphere, where the vapours are *suddenly condensed into heavy rains*, by cold currents from the poles.

428. *Why does the greatest quantity of rain fall at the equator?*

Because the *hot air* absorbs a large amount of vapour, and as the atmosphere is usually calm, there is an absence of currents, by which the saturated air would be removed. In this, which is called "*the Region of Calms,*" rain falls almost daily.

429. *Why are some parts of the earth rainless?*

"Thou, O God, didst send a plentiful rain, whereby thou didst confirm thine inheritance, when it was weary."—PSALM LXVIII.

Because, being situated in tropical or torrid latitudes, and at a distance from the ocean, the atmosphere above them is always in a *dry state.*

430. *When is air said to be saturated with vapour?*

When it cannot take up *a larger quantity* than that which it already holds.

When common salt is dissolved in water, until the water can take up no more, the water is then said to be *saturated with salt.*

431. *What proportion of water is air capable of sustaining in the form of vapour?*

The amount of water held in suspension by the air averages the following proportion: one thousand *cubic feet of air* contain as much vapour as, were it condensed to water, would yield about *two fifths of a pint.*

But *one thousand cubic feet of air* are capable of holding *half-a-pint of water;* and this may be regarded as the *point of saturation.*

Thus, in a room ten feet square and ten feet high, the air, *at the point of saturation,* would hold in the form of vapour, *half-a-pint of water.* It must not be forgotten, however, that the point of saturation necessarily varies with the *temperature of the air.*

432. *Why are cloudy days and nights not always wet?*

Because the air has not reached the state of *saturation.*

433. *Why does rain purify the air?*

Because it produces motion in the particles of the air, by which they are *intermixed.* And it precipitates noxious *vapours,* and cleanses the face of the earth from *unhealthy accumulations.*

434. *Why are mountainous localities more rainy than flat ones?*

Because the mountains *attract the clouds;* and because the clouds that are flying low are borne against the sides of the mountains and directed upwards, where they meet with *cold currents of air.*

435 *Why does more rain fall by night than by day?*

Because by night the temperature of the air, heated during the day, falls to that degree which condenses *its vapours into rain.*

436. *Why do bunches of dried sea-weed indicate the probability of coming rain?*

Because they readily imbibe moisture, and when they become soft and damp they show that the air is *approaching the point of saturation.*

437. *Why does the weather-toy, called the "weather-cock," foretell the probability of rain?*

Because it is made with a piece of cat-gut which swells with moisture, and as it swells, ***shrinks.*** The cat-gut is so applied that when it ***shrinks,*** it turns a rod which sends the ***man*** out of the house, and when it ***dries*** it sends the ***woman out.*** Therefore, when the ***man*** appears, it is a sign of ***wet,*** and when the ***woman*** appears it is a sign of ***dry weather.***

There is another toy, called the Capuchin, which is made upon the same principle. The figure lifts a hood over its head when wet is approaching, and takes it off when the weather is becoming dry. In this case, a piece of cat-gut is also employed. Various weather-toys may be made upon this principle—among others, a little umbrella, which will open on the approach of wet, and close on the return of fine weather.

A gentleman once made a wooden horse, which he declared should of itself walk across a room, without machinery of any kind. The assertion was discredited; but the horse was placed in a room close to the wall on one side. The room was locked, and otherwise fastened, so that no one could interfere with the experiment. After a time the door was opened, and it was found that the horse had actually crossed the floor, and stood on the opposite side. The horse was made from wood of a peculiar kind, liable to great expansion in wet weather, and cut in a manner to produce the greatest elongation. The fore hoofs were so made that where they were set they would remain, so that the contracting parts should draw up from behind. It is easy to understand how, in this way, the wooden horse crossed the apartment.

438. ***Why does ladies' hair drop out of curl upon the approach of damp weather?***

Because the hair ***absorbs moisture,*** which causes its spirals to relax and unfold.

439. ***Why is it said in mountainous countries that rain is***

"Hast thou entered into the treasures of the snow; or hast thou seen the treasures of the hail."—JOB XXXVIII.

coming, because the mountains are "putting their night-caps on?"

Because the clouds descend when they are *heavy with vapour*, and being attracted to the mountain tops they are said to "*cap the mountains.*"

CHAPTER XXI.

440. *What is snow?*

Snow is *congealed vapour*, which would have formed ***rain;*** but, through the coldness of the air, has been *frozen* in its descent into *crystaline forms.* (Fig. 1.)

441. *Why is snow white?*

Because it reflects all the component rays of *light.*

442. *Why is snow said to be warm, while white garments are worn for coolness?*

Snow is *warm* by virtue of its light and woolly texture. But it is also warm on account of its *whiteness;* for, had it been *black*, it would have *absorbed the heat of the sun*, which would have *thawed the snow.* Instead of which, it *reflects heat;* and the reflected heat *falls upon* bodies above the snow, while the *warmth of the earth* is preserved *beneath it. White clothing* is *cool*, because it reflects *from* the body of the wearer the heat of the sun. *White snow* is *warm*, because it *reflects the sun's heat upon bodies.*

There are few persons but have felt the effect of the sun's rays *reflected* by the white snow on a clear wintry day. And, as regards the warmth of snow towards the earth, by preventing the radiation of heat, it has been found that a thermometer buried four inches deep in snow has shown a temperature of *nine degrees* higher than at the surface.

443. *Why are lofty mountains always covered with snow?*

Because the *upper regions* of the atmosphere are *intensely cold.*

444. *Why are the upper regions of the atmosphere intensely cold?*

Because the *atmosphere* retains but *little of the heat of the sun's*

"He causeth the vapours to ascend from the ends of the earth: he maketh lightnings for the rain; he bringeth the wind out of his treasuries."—Ps. xxxv

rays as they pass to the earth. Because at high altitudes the air is *greatly rarefied*. And because the *radiation of heat from the earth* does not materially affect such *high regions.*

445. *What is meant by the snow line?*

The *snow line* is the estimated altitude in *all countries* where *snow would be formed.* Even at the equator, at an altitude of 15,000 to 16,000 feet from the level of the sea, snow is found upon the mountain summits, where it perpetually lies. As we proceed north or south from the equator the *snow line lessens in altitude.* Had we in England a mountain 6,000 feet high, it would be perpetually *crowned with snow.*

446. *Why do we hear of red snow?*

Red snow is the name given to the snow in the arctic regions upon which a minute vegetable (probably the *Protoccus nivalis*) grows, imparting to the snow a red colour. Recent microscopic investigations have shown it to consist of a minute vegetable cell, which secretes a red colouring matter.

Snow is found to be of greater import nce to man than is generally supposed. But, although in this country we are enabled to recognise the hand of Providence in the gift, there are latitudes wherein the blessing thus conferred is more deeply felt. In such countries as Canada, Sweden, and Russia, the falling of snow is looked for with glad anticipations, quite equalling those which herald the "harvest-home" of England, or the "vintage" of France. No sooner is the ground covered with snow, than cranky old vehicles that had been jolting over rough roads, and sticking fast in deep ruts of mud, are wheeled aside, and swift sledges take their place. Towns distant from each other find an easy mode of communication; the markets are enlivened, and trade thrives. Snow supplies a kind of railroad, covering the entire face of the country, and sledges glide over it, almost with the speed of the locomotive.

447. *What is sleet?*

Sleet is snow which, in falling, has met with a *warmer current of air* than that in which it congealed. It therefore partially melts and forms a kind of *wet snow.*

448. *What is hail?*

Hail is also the *frozen moisture of the clouds.* It is probably formed by *rain drops* in their descent to the earth, meeting with an *exceedingly cold current of air* by which they become ***suddenly*** *frozen into hard masses.*

"If the clouds be full of rain, they shall empty themselves upon the earth."—ECCLES. XI.

It is also supposed that the *electrical* state of the air and of the clouds influences the formation of *hail.*

449. *Why is it supposed that the electrical state of the air and the clouds affects the formation of hail?*

Because hail is more common in the *summer* than at other seasons, and is frequently attended by storms of *thunder and lightning.*

450. *Why do hail-storms most frequently occur by day?*

Because the clouds, being charged with vapour to saturation, favour the formation of hail by *sudden* electrical or atmospheric changes. In the gradual cooling of night, the clouds would expend themselves in rain.

Astonishing facts respecting hail-storms are upon record. In 1719 there fell at Kremo, hailstones weighing six pounds. In 1828 there was a fall of ice at Horsley, in Staffordshire, some of the pieces of which were three inches long, by one inch broad; and other solid pieces were about three inches in circumference. Hail storms are most frequent in June and July, and least frequent in April and October. Hail clouds float much lower in the sky than other clouds; their edges are marked by frequent heavy folds; and their lower edges are streaked with white, the other portions being massive and black. (Fig. 10.)

CHAPTER XXII.

451. *What is light?*

Light, according to Newton, is the effect of luminous particles which dart from the surfaces of bodies in all directions. According to this theory, the solar light which we receive would *depart from the sun and travel to the earth.*

According to Huyghens, light is caused by an *infinitely elastic ether, diffused through all space.* This ether, existing everywhere, is *excited into waves, or vibrations, by the luminous body.*

The theory of light is so undetermined that neither the views of Newton, and those of Huyghens, can be said to be exclusively adopted. Writers upon natural philosophy seize hold of either or both of those theories, as they present themselves more or less favourably in the explanation of natural phenomena. In "*The Reason Why,*" as we have to speak of the *effects* of light rather than of its *cause,* we shall avoid, as far as possible, the doubtful points. But let no one be discouraged by the fact that the theory of light, as, indeed, of all the imponderable agents, is imperfectly understood. Rather let us rejoice that there are vast fields of discovery yet to be explored; and that light, the most glorious and inspiring element in nature, invites us from the sun, the moon, and the stars, and from the face of every green leaf and variegated flower, to search out the wonders of its nature, and further to exemplify the goodness and wisdom of God.

452. *What is the distance of the sun from the earth?*

Ninety five millions of miles.

453. *At what rate of velocity does light travel?*

At the rate of 192,000 miles in a *second*, through our *atmosphere;* and 192,500 miles in a *second* through a *vacuum.*

454. *How long does light take to travel from the sun to the earth?*

Eight minutes and thirteen seconds.

455. *What is the constitution of the sun?*

It is a spherical body, 1,384,472 times larger than the earth.

456. *From what does the luminosity of the sun arise?*

From a luminous atmosphere, or, as M. Arago named it, *photosphere,* which completely surrounds the body of the sun, and which is probably *burning with great intensity.*

457. *What are the minor sources of light?*

Light may be produced by *chemical action,* by *electricity,* and by *phosphoresence,* in the latter of which various agencies unite.

458. *What is a ray of light?*

A *ray* of light is the *smallest portion* of light which we can recognise.

459. *What is a medium?*

A *medium* is a body which affords *a passage for the rays* of light.

460. *What is a beam of light?*

A *beam* of light is a *group of parallel rays.*

461 *What is a pencil of light?*

"And God saw the light, that it was good: and God divided the light from the darkness."—GEN. I.

A *pencil* of light is a body of rays which *come from or move towards a point.*

462. *What is the radiant point?*

The *radiant point* is that *from which diverging rays of light are emitted.*

463. *What is the focus?*

The *focus* is the point to which *converging rays are directed.*

Diverging, starting from a point, and separating. *Converging*, drawing together towards a point.

464. *What is the constitution of a ray of light?*

A ray of *white light*, as we receive it from the sun, is composed of *a number of elementary rays*, which, with the aid of a triangular piece of glass, called a *prism*, may be separated, and will produce under refraction the following colours :—

1. An *extreme red* ray—a mixture of *red* and *blue*, the red *predominating*.
2. *Red.*
3. *Orange*—red passing into and combining with yellow.
4. *Yellow*—the most luminous of all the rays.
5. *Green*—yellow passing into and combining with the blue.
6. *Blue.*
7. *Indigo*—a dark and intense blue.
8. *Violet*—blue mingled with red.
9. *Lavender grey*—a neutral tint.
10. Rays called *fluorescent*, which are either of a *pure silvery blue*, or a *delicate green.*

465. *Why is a ray of light, which contains these elementary rays, white?*

Because the colour of light is governed by the *rapidity of the vibrations of the ether-waves.* When a ray of light is refracted by, or transmitted through a body, its *vibrations are frequently disturbed and altered*, and thus a *different impression* is made upon the *eye.*

Light which gives 37,640 vibrations in *an inch*, or 458,000,000,000,000 in a *second of time*, produces that sensation upon the eye which makes the object that directs the vibrations appear *red.* *Yellow* light requires 44,000 vibrations *in an inch,* and 535,000,000,000,000 in a *second of time.* And the other colours enumerated (*see* 464) all require different *velocities of vibration* to produce the colours by which they are distinguished.

Accepting the theory of vibrations, and applying it to the elucidation of the phenomena of light—it is unnecessary, we think, to believe that a ray of *white* light *contains* rays in *a state of colour.* It is said that if we divide a circular surface into parts, and paint the various colours in the order and proportions in which they occur in the refracted ray, and then spin the circle with great velocity, the colours will blend and appear *white.* But such is not the case; the result is in some degree an illusion, arising out of the sudden removal of the impression made upon the eye by the colours; and if a piece of white paper be held by the side of the coloured circle in motion, the latter will be found to be *grey.* When it is remembered that in colouring a white surface with thin colours, the white materially qualifies the colours, it must be admitted that the experiment fails to support the assertion that the colours of the spectrum produce white. But there can be no difficulty in understanding that a ray of light undergoing *refraction,* becomes divided into minor rays, which *differing in their degrees of refrangibility,* vary also in the *velocity of their vibrations,* and produce the several sensations of colour.

466. *Why is a substance white?*

Because it reflects the light that falls upon it ***without altering its vibrations.***

467. *Why is a substance black?*

Because it ***absorbs the light*** and ***puts an end to the vibrations.***

468. *Why is the rose red?*

Because it imparts to the light that falls upon it that ***change in its vibratory condition,*** which produces on our eyes the *sensation* of *redness.*

469. *Why is the lily white?*

Because it reflects the light without altering its vibrations.

470. *Why is the primrose yellow?*

Because, though it receives white light, it alters its vibrations to 44,000 in an inch, and 535,000,000,000,000 in a second, and this is the velocity of vibration which produces upon the eye a ***sensation*** *of yellow.*

"But if thine eye be evil, thy whole body shall be full of darkness. If therefore the light that is in thee be darkness, how great is that darkness."– MATT. v.

471. *Why are there so many varieties of colour and tint in the various objects in nature?*

Because every surface has a peculiar constitution, or atomic condition, *by which the light falling upon it is influenced.* In tropical climates, where the brightness of the sun is the most intense, there the colours of natural objects are the richest; the foliage is of the darkest green; the flowers and fruits present the brightest hues; and the plumage of the birds is of the most gaudy description. In the temperate climates these features are more subdued, still bearing relation to the degree of light. And at a certain depth of the ocean, where light penetrates only in a slight degree, the objects that abound are nearly colourless.

It has been held by many philosophers (and the theory is so far conclusive that it cannot be dispensed with) that there is an analogy between the vibratory causes of *sound*, and the vibratory causes of *colour*. Any one who has seen an Æolian harp, and listened to the wild notes of its music, will be aware that the wires of the harp are swept by accidental currents of air; that when those currents have been strong, the notes of the harp have been raised to the highest pitch, and as the intensity of the currents has fallen, the musical sounds have deepened and softened, until, with melodious sighing, they have died away. No finger has touched the strings; no musical genius has presided at the harp to wake its inspiring sounds; but the vibration imparted to the air, as it swept the wires, has alone produced the chromatic sounds that have charmed the listener. If, then, the varied vibrations of the *air* are capable of imparting dissimilar sensations of *sounds* to the *ear*, is it not only possible, but probable, that the different vibrations of *light* may impart the various sensations of *colours* to the *eye?*

CHAPTER XXIII.

472. *What is the refraction of light?*

When rays of light fall *obliquely* upon the surface of any *transparent medium*, they are slightly diverted from their course. This alteration of the course of the rays is called *refraction*, and the degree of refraction is influenced by the difference between the *densities* of the mediums *through which light is transmitted.*

"Let your light so shine before men, that they may see your good works, and glorify your father which is in heaven."—MATT. V.

473. *If a ray of light falls in a straight line upon a transparent surface, is it then refracted?*

In that case the ray pursues its course—*there is no refraction.*

474. *Is the direction in which the rays are bent, or refracted, influenced by the relative densities of the media?*

A ray of light falling slantingly upon a *window,* in passing through it is slightly brought to the *perpendicular;* and if it then falls upon the surface of water, it is still further brought to the perpendicular in *passing through the water.*

475. *Is light refracted in passing from a dense medium to a thinner one?*

It is; but the *direction of the refraction* is just the opposite to the instance just given; a ray of light passing through *water* into *air,* does not take a more *perpendicular course,* but becomes more *oblique.*

Fig. 11.

476. *Why, if a rod or a spoon be set in an empty basin, will it appear straight, or of its usual shape?*

Because the rays of light that are reflected from it all pass through the same medium, the *air.*

477. *Why if water be poured into the basin will the rod or spoon appear bent?*

Because the rays of light that pass through the *water* are *reflected in a differen' degree* to those that pass through the air

'Evening, and morning, and at noon, will I pray, and cry aloud; and he shall hear my voice."—PSALM LV.

Place in the bottom of an empty basin (Fig. 11.) a shilling; then stand in such a position at the point B that the line of sight, over the edge of the basin, just excludes the shilling from view. Then request some one to pour water into the basin, until it is filled to C (Fig. 12.), keeping your eye fixed upon the spot. The shilling will gradually appear, and will soon come entirely in view. Not only will the shilling be brought in view, but also portions of the basin before concealed. This is owing to the rays of light passing from the bottom through the water in a direction *more perpendicular* than they would have done through the air; but on leaving the water they become more *oblique*, and hence they convey the image of the shilling ***over the edge of the basin***, **which otherwise would have obstructed the view.**

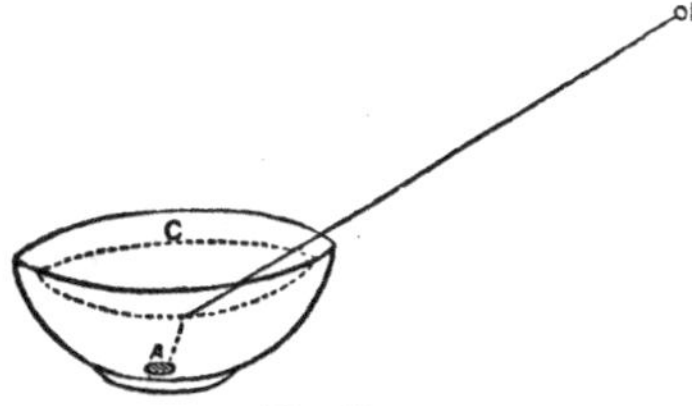

Fig. 12.

478. *Why is it that in cloudy and showery days we see the sun's rays bursting through the clouds in different directions?*

Because, in passing through clouds of *different densities* the rays are *bent out of their course.*

479. *Why is the apparent depth of water always deceptive?*

Because the light reflected from the objects at the bottom is *refracted* as it leaves the water.

480. *How much deeper is water than it appears to be?*

About *one-third.* A person bathing, and being unable to swim, should calculate before jumping into the water, that if it *looks two feet deep*, it is quite *three feet.*

481. *Why can we seldom at the first attempt touch anything lying at the bottom of the water with a stick?*

Because we do not allow for the *different refractive powers* of water and of air

482. *Why do we see the sun before sun-rise, and after sun-set?*

Because of the refractive effects of the atmosphere. Rays of light, passing obliquely from the sun through the air to the earth, are refracted three or four times by the varying density of the medium. Each refraction bends the rays towards the *perpendicular;* and hence we see the sun *before it rises* and *after it sets.*

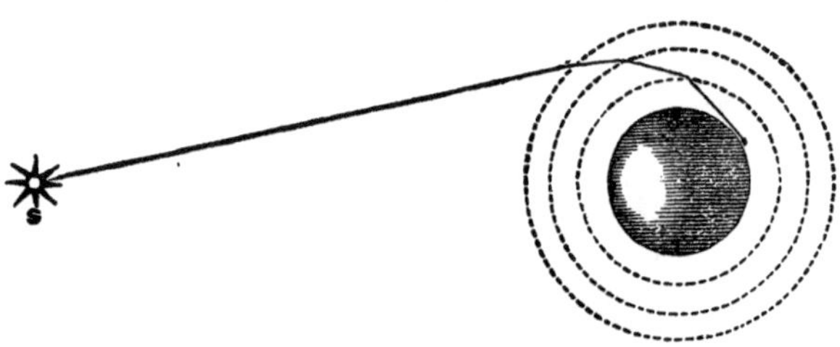

Fig. 13.—DIAGRAM EXHIBITING THE REFRACTION OF THE SUN'S RAYS IN PASSING THROUGH THE ATMOSPHERE.

483. *Why do figures, viewed through the hot air proceeding from furnaces, and from lime-kilns, appear distorted and tremulous?*

Because the ever varying density of the air which is flying away in hot currents, and succeeded by cold, *constantly changes the refractive power* of the medium through which the figures are viewed.

484. *Why do the stars twinkle?*

Because their light reaches us through *variously heated and moving currents of air.* In this case the earth is the *kiln,* and the *stars* the *object* that is *viewed through the refractive medium.*

485. *Why does much twinkling of the stars foretell bad weather?*

Because it denotes that there are *various aerial currents* of different temperatures and densities, producing *atmospheric disturbance.*

"And it shall come to pass, when I bring a cloud over the earth, that the bow shall be seen in the cloud."—GENESIS IX.

486. *What causes the rainbow?*

The *refraction* of the sun's rays by the *falling rain.*

487. *Why does the rainbow exhibit various colours?*

The colours belong to the *elementary rays of light;* and these rays having *different degrees of refrangibility,* some of them are bent more than others; they are therefore separated into *distinct rays of different colours.*

488. *Why are there sometimes two rainbows?*

Because the rays of *refracted* light, reflected upon other drops of rain, are *again* refracted, and then *reflected again,* forming a secondary bow.

489. *Why are the colours of the secondary bow arrayed in the reverse order of the primary bow?*

Because the secondary bow is *a reflection* of the primary bow, and, like all reflections, is reversed.

490. *Why are reflections reversed?*

Because those rays which *first reach* the reflecting surface are the *first returned.* If you hold your open hand towards the looking-glass, the light passing from the point of your finger will reach the reflector and be returned before the rays that pass from the back parts of the hand. Hence the image of the hand will present the reflection of the finger point towards the point of the finger.

491. *Why are the colours of the secondary rainbow fainter than those of the primary?*

Because they are derived from the *refraction and reflection* of rays which have *already* been refracted and reflected, and thereby *their intensity has been diminished.*

492. *What is a lunar rainbow?*

A *lunar rainbow* is caused by the light of the *moon,* in the same manner as the *solar rainbow* is caused by the light of the sun

"I am come a light into the world, that whosoever believeth in me should not abide in darkness."—JOHN XIII.

493. *Why is the lunar rainbow fainter than a solar rainbow?*

Because the *light of the moon* is the *reflected light of the sun*, and is therefore *less intense.*

494. *What is a halo?*

A halo is a *luminous ring*, which forms between the eye of the observer and a luminous body.

Haloes may appear around the disc of the sun, moon, or stars. But in this country the *lunar* haloes are the most remarkable and frequent.

495. *What is the cause of the luminous ring?*

The *refraction of light* as it passes through an intervening *cloud*, or a stratum of *moist* and *cold air.*

496. *Why are haloes sometimes large and at other times small?*

Because they are sometimes formed *very high* in the atmosphere, at other times *very low.* Being high, and farther removed from the spectator, and nearer the source of light, they appear *smaller;* while the nearer they are, the *larger they appear.*

497. *Why do haloes foretell wet weather?*

Because they show that there is a great amount of atmospheric moisture, which will probably form *rain.*

498. *Why do glass lüstres and chandeliers exhibit "rainbow colours"?*

Because they *refract the rays of light* in the same manner as the rain drops.

499. *Why does a soap bubble show the prismatic colours?*

Because, like a large rain drop, it *refracts the rays of light*, and shows the elementary rays.

500. *What causes the rich tints displayed by "mother-of-pearl?"*

The *refraction of the light* that falls upon the surface of the pearl.

"Light is sown for the righteous, and gladness for the upright in heart."—
PSALM XCVII.

501. *What causes the brilliant colours of the diamond?*

The *refraction* of the rays of light by the various *facets* of the diamond.

The refraction of light, and the production of prismatic colours, surrounds us with most interesting phenomena. The laundress, whose active labours raise over the wash-tub a soapy froth, performs inadvertently one of the most delicate operations of chemistry—the chemistry of the imponderable agents—and the result of her manipulations manifests itself in the delicate colours that dance like a fairy light over the glassy films that follow the motion of her arms. The laughing child, throwing a bubble from the bowl of a tobacco pipe into the air performs the same experiment, and produces a result such as that which filled the philosophic Newton with unbounded joy. The foam of the sea shore, the plumage of birds, the various films that float upon the surface of waters, the delicate tints of flowers, and the rich hues of luscious fruits, all combine to remind us, that every ray of light comes like an angelic artist sent from heaven, bearing upon his palette the most celestial tints, with which to beautify the earth, and show the illimitable glory of God.

CHAPTER XXIV.

502. *What is the difference between the refraction and the reflection of light?*

Refraction is the deviation of rays of light from their course through the interference of a *different* medium; *reflection* is the return of rays of light which, having fallen upon a surface, are repelled by it.

503. *What is the radiation of light?*

The *radiation* of light is its *emission in rays* from the surface of a *luminous body*.

504. *Do all bodies radiate light?*

All bodies radiate light; but those that are not in themselves primary sources of light, are said to *reflect it*.

505. *Do black bodies reflect any light?*

Black bodies *absorb* the light that falls upon them. But they reflect a *very small* degree of light.

506 *Why is glass transparent?*

"As in water face answereth to face, so the heart of man to man. —
PROVERBS XXVII.

Because its atoms are so arranged that they allow the vibrations of light to continue through their substance.

507. *Does glass obstruct the passage of any portion of light?*

Glass *reflects* (sends back) a very small portion of light. This may be observed by holding a piece of paper, or a hand, a few inches from a window, when a faint reflection of it will be visible. Probably the small amount of light *reflected by transparent glass, which gives a passage to the greater part of the rays*, may serve to illustrate the small amount of light reflected from *black surfaces*, which *absorbs the greater portion of light.*

Instead of a piece of white paper, hold a piece of *black cloth* two or three inches from the window-pane, and you will have two reflections so weak that the image of the cloth will be almost lost. The first reflection is that of the very small amount of light from the black surface on to the glass, and the second reflection is that of the inconceivably small amount returned by the glass, and by which the faint image of the black cloth is produced. But put the black cloth outside of the window-pane, and then hold an object before them, and you will find that the *two weak reflectors, acting together*, produce an improved image, or reflection.

508. *Why, if a book is held between a candle-light and the wall, does a shadow fall upon the wall?*

Because the rays of light are *intercepted* by the book.

509. *Why do the rays pass over the edges of the book in a direct line with the flame of the candle?*

Because light always travels in *straight lines.*

510. *Why is there some amount of light even where shadows fall?*

Because, *as all objects reflect light*, some of them throw their light into the field of the shadow.

511. *Why are some substances opaque to light?*

Because the arrangement of their particles will not admit of the *vibrations of the luminous ether* passing through them.

Opaque—impervious to rays of light.

512. *Why do we see our faces reflected in mirrors?*

Because the rays of light from our faces are *reflected* by the surface of the *quicksilver* at the back of the glass.

513. *Why does the quicksilver reflect the rays of light?*

Because, being *densely opaque to light,* and presenting also a bright surface, it is a good reflector, and it *throws back the whole of the rays.*

514. *What has the glass to do with the reflection?*

The glass has *nothing to do with the reflection,* except that it affords a field upon which the reflecting surface of the quicksilver is spread; and it keeps the air and dirt from *dulling the quicksilver.*

The parts of a mirror from which the quicksilver is rubbed away give no reflection that could assist the reflecting power of the quicksilver. That the surface of the glass does not reflect the image, is shown by the fact, that if you put the point of any object against the glass, the thickness between the point and the place where the reflection of it begins, will *show the exact thickness of the glass.*

515. *Why does a compound mirror (a multiplying mirror) exhibit a large number of images of one object.*

Because all objects reflect rays of light in *every direction,* and therefore the different mirrors, being at *various angles,* receive *each a reflection* of the same object.

516. *Why does a window-pane appear to be a better reflector by candle-light than by day-light?*

The reflecting power of glass is precisely the same by night as by day, and is always very feeble. But it appears to be greater by night, *because the surrounding darkness increases the apparent strength of the reflection.*

517. *How do we know that objects reflect light in every direction?*

Because if we *prick a hole in a card with a pin,* and then look through that small hole upon a *landscape,* we can see some miles of country, and some thousands of objects; every part of every object throughout the whole scene, must have sent rays of light to the small hole pricked in the card.

'Such knowledge is too wonderful for me; it is high, I cannot attain unto it."—
PSALM CXXXIX.

At one extremity of the landscape, viewed through the hole in the card, there may be a forest of trees; in the distance there may be hills bathed in golden light, and overhung with glittering clouds; in the mid-distance there may be a river winding its course along, as though it loved the earth through which it ran, and wished, by wandering to and fro, to refresh the thirsty soil; in the foreground may be a church, covered by a million ivy leaves; and grouping towards the sacred edifice may be hundreds of intending worshippers, old and young, rich and poor; flowers may adorn the path-ways, and butterflies spangle the air with their beauties: yet every one of those objects—the forest, the hills, the clouds, the river, the church, the ivy, the people, the flowers, the butterflies—must have sent rays of light, which found their way through the little hole in the card, and entered to paint the picture upon the curtain of the eye.

This is one of the most striking instances that can be afforded of the wonderful properties of light, and of the infinitude of those luminous rays that attend the majestic rising of the sun. Not only does light fly from the grand "ruler of the day" with a velocity which is a million and a half times greater than the speed of a cannon-ball, but it darts from every reflecting surface with a like velocity, and reaches the tender structure of the eye so gently that, as it falls upon the little curtain of nerves which is there spread to receive it, it imparts the most pleasing sensations, and tells its story of the outer world with a minuteness of detail, and a holiness of truth. Philosophers once sought to *weigh the sunbeam;* they constructed a most delicate balance, and suddenly let in upon it a beam of light; the lever of the balance was so delicately hung that the fluttering of a fly would have disturbed it. Everything prepared, the grave men took their places, and with keen eyes watched the result. The sunbeam that was to decide the experiment had left the sun eight minutes prior to pass the ordeal. It had flown through ninety-five millions of miles of space in that short measure of time, and it shot upon the balance with unabated velocity: but the lever moved not, and the philosophers were mute.

CHAPTER XXV.

518. ***Why, when we move before a mirror, does the image draw near to the reflecting surface as we draw near to it, and retire when we retire?***

Because the lines and angles of *reflection* are always equal to the lines and angles of *incidence.*

519. ***What is the line of incidence?***

If a person stands in a direct line before a mirror, the line through which the light travels from him to the mirror is *the line of incidence.*

Incidence—falling on.

"Blessed be the Lord, who daily loadeth us with benefits, even the God of our salvation."—PSALM LXVIII.

520. *What is the line of reflection?*

The *line of reflection* is the line in which the rays of light are returned from the image formed in the glass to the eye of the observer.

Reflection—a turning back.

521. *What is the angle of incidence?*

The *angle* of incidence is the angle which rays of light, falling on a reflecting surface, make with a line perpendicular to that surface.

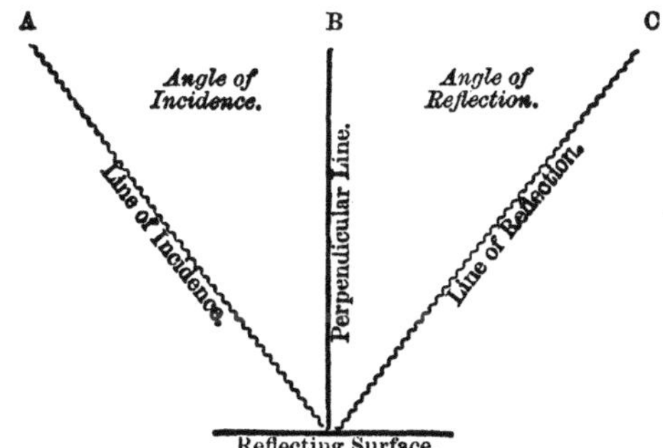

Fig. 14.—EXPLAINING THE LINES AND ANGLES OF INCIDENCE AND OF REFLECTION.

522. *What is the angle of reflection?*

The *angle* of reflection is the angle which is formed by the returning rays of light, and a line perpendicular to the reflecting surface. It is always *equivalent* to the angle of incidence.

Take a marble and roll it across the floor, so that it shall strike the wainscot obliquely. Let A in the diagram represent the point from which the marble is sent. The marble will not return to the hand, nor will it travel to the line B, but will bound off, or be *reflected*, to C. Now B is an imaginary line, *perpendicular to the reflecting surface;* and it will be found that the path described by the marble in *rolling to the surface and rebounding from it,* form, with the line B, two angles that are *equal.* These represent the angles of *incidence* and of *reflection,* and explain why the reflection of a person standing at A before a mirror, would be seen by another person standing at C. This simple law in optics explains a great many interesting phenomena, and therefore it should be clearly impressed upon the memory.

6

"And God made two great lights; the greater light to rule the day, and the lesser light to rule the night: he made the stars also."—GEN. I.

523. *Why do windows reflect the sun in the evening?*

Because the eye of the observer is in the *line of the reflection.*

524. *Why do windows not reflect the sun at noon?*

They do, but our eyes are not then in the *line of the reflection.*

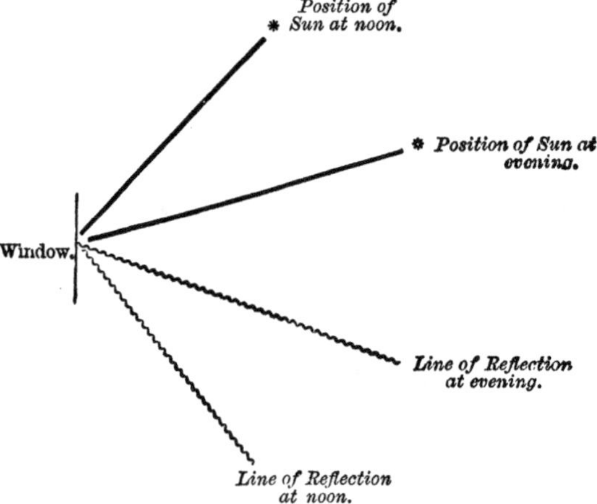

Fig. 15.—SHOWING THE LINES OF INCIDENCE AND REFLECTION OF THE SUN'S RAYS AT NOON AND AT EVENING

It is obvious from the foregoing diagram that the evening rays of reflection fall upon the eyes of spectators, while the reflections at noon are so perpendicular that they are lost.

525. *Why do the sun and moon appear smaller when near the meridian, than when near the horizon?*

Because, when near the horizon, they are brought into *comparison with the sizes of terrestrial objects*; but when near the meridian they occupy the centre of a vast field of sky, and as there are no objects of comparison surrounding them, they *appear smaller.*

This is one "Reason Why," assigned by some observers. But there is also another reason to be found in the fact that, when the sun or moon is near the horizon, we view it through a *greater depth of atmosphere* than we do when at

" There is no darkness nor shadow of death, where the workers of iniquity may hide themselves."—JOB XXXIV.

the meridian. (*See* Fig. 13.) A straight line passed upward through the air would not be so long as that which passes to S. Consequently, as the air is generally impregnated with moisture, at the time when these effects are observed, the rays of light are caused to diverge more, and the disc of the sun or moon *appears magnified*. Probably both of these reasons contribute to the effect. This latter reason also explains why the disc of the sun or moon may sometimes appear *oval* in shape, the lower stratum of air being more loaded with moisture than that through which we view the upper part of the disc.

526. *Why do our shadows lengthen as the sun goes down?*

Because light travels only in ***straight lines***, and as the sun descends, the direction of his rays becomes more ***oblique***, thereby causing longer shadows.

527. *What is the cause of the optical illusions frequently observed in nature?*

There are various kinds of natural optical illusions :—

The *mirage*, in which landscapes are seen reflected in burning sands.

The *fata morgana*, in which two or three reflections of objects occur at the same time.

The *ærial spectra*, or ærial reflections, &c.

Fig. 16.—ILLUSTRATING THE APPEARANCE OF PHANTOM SHIPS.

The optical illusions above enumerated owe their origin to various atmospheric conditions, in which *refractions* and *reflections* are *multiplied* by the different densities of atmospheric layers. They chiefly occur in hot countries, where, from the varying effects of heat, the conditions of atmospheric refraction and reflection frequently prevail in their highest degree.

528. *Why do we have twilight mornings and evenings?*

Because the coming and the departing rays of the sun are *refracted* and *reflected* by the upper portions of the atmosphere. (*See* Fig. 13.)

529. *How long before the sun appears above the horizon does the reflection of his light reach us?*

The time *varies* with the refracting and reflecting power of the atmosphere, from *twenty minutes* to *sixty minutes*. But the sun's position is usually *eighteen degrees* below the horizon when twilight begins or ends.

530. *Why is the sky blue?*

The white light of the sun falls upon the earth without change; it is then reflected back by the earth, and as it passes through the atmosphere portions of it are again returned to us, and this double reflection produces a *polarised* condition of light which imparts to vision the sensation of a *delicate blue*. (*See* 549.)

531. *Why do the clouds appear white?*

Because they reflect back to us the solar beam *unchanged.*

532. *Why does the sky appear red at sunset?*

Because the light vapours of the air, which are condensed as the sun sets, refract the rays of light, and produce red rays. The refraction which produces *red* requires only a *moderate degree of density*.

533. *Why do the clouds sometimes appear yellow?*

Because there is a larger amount of vapour in the air, which produces a different degree of refraction, *resulting in yellow.*

534. *Why does a yellow sunset foretell wet weather?*

Because it shows that the air is heavy with vapours. The refraction that produces *yellow* requires a greater degree of density.

535. *Why does a red sunset foretell fine weather?*

Because the redness shows that the vapours in the air *towards the West*, or wet quarter, are *light*, as is evidenced by the degree of refraction of the sun's rays.

536. *Why does a red sunrise foretell wet?*

Because it shows that *towards the East*, or dry quarter, the air is charged with vapour, and therefore probably at other points the air has reached *saturation*.

537. *Why does a grey sunrise foretell a dry day?*

Because it shows that the vapours in the air are *not* very dense.

538. *Why is "a rainbow in the morning the shepherd's warning?"*

Because it shows that *in the West*, or wet quarter, the air is *saturated* to the rain point.

539. *Why is "a rainbow at night the shepherd's delight?"*

Because it shows that the *rain is falling in the East*, and as that is a dry quarter, it will *soon be over*. Rainbows are always seen in opposition to the sun.

CHAPTER XXVI.

540. *What is the difference between light and heat?*

The most obvious distinction is, that light acts upon *vision*, and heat upon *sensation*, or feeling.

Another distinction is, that *heat expands all bodies*, and alters their atomic condition; while *light*, though usually attended by heat, does not display the same expansive force, but produces various effects which are *peculiar to itself*.

"Ye are the light of the world. A city that is set on a hill cannot be hid."—
MATTHEW V.

541. *Are light and heat combined in the solar ray?*

Yes. A ray of light, as well as containing elementary rays that produce colours under refraction, contains also *chemical rays,* and *heat rays.*

542. *How do we know that light and heat are separate elements?*

Because we have *heat rays,* as from dark hot iron, from various chemical actions, and from friction, which are *unattended by the development of light.* And we have light, or luminosity, such as that of *phosphoresence,* which is unaccompanied by any appreciable degree of heat.

But, besides this confirmation, further proof is afforded by the fact, that in passing rays of solar light through media that are *transparent to heat,* but not to *light,* the heat rays may be *separated* from the luminous rays, and *vice versa.*

Black glass, and black mica, which are nearly *opaque to light,* are *transparent to heat* to the extent of ninety degrees out of a hundred. While pale green glass, coloured by oxide of copper, and covered with a coating of water, or a thin coating of alum, will be perfectly *transparent to light,* but will be almost quite *opaque to heat.* These remarks apply, in a greater or less degree, to various other substances.

543. *In what respects are light and heat similar?*

Both heat and light have been referred to minute vibratory motions which occur, under exciting causes, in a very subtile elastic medium.

They are both united in the sun's rays.

They are both subject to laws of absorption, radiation, reflection, and refraction.

They are both essential to life, whether animal or vegetable.

Both may be developed in their greatest intensity by electricity.

They are both imponderable.

544. *In what respects are light and heat dissimilar?*

Heat frequently exists without light.

Light is usually attended with heat.

Light may be instantly extinguished, but

Heat can only be more gradually reduced, by diffusion.

The solar rays deliver heat to the earth by day, and the heat remains with the earth when the light has departed.

Heat diffuses itself in all directions.

Light travels only in straight lines.

The colours that absorb and radiate both light and heat do not act in the same degree upon them both. Black, which does not radiate light, is a good *radiator of heat*, &c., &c.

The oxy-hydrogen *light* emits a most intense heat, but glass which will transmit the rays of light, will afford no passage to the rays of the *heat*.

Heat is latent in all bodies, but no satisfactory proof has been found that light is latent in substances.

These are only a few of the analogies and distinctions that exist between the two mysterious agents, light and heat. But they are sufficient to supply the starting points of investigation.

The importance of the heat that attends the solar rays may be illustrated by the experiments performed a few years ago, by Mr. Baker, of Fleet-street London, who made a large burning lens, three feet and a half in diameter, and employed another lens to reduce the rays of the first to a focus of half an inch in diameter. The heat produced was so great that iron plates, gold, and stones were *instantly melted;* and sulphur, pitch, and resinous bodies, *were melted under water.*

545. *What is the point of heat at which bodies become luminous?*

The point of heat at which the eye begins to discover luminosity has been estimated at 1,000 deg.

546. *What is the velocity of artificial light?*

The light of a fire, or of a candle, or gas, travels with the same velocity as the light of the sun,—a velocity which would convey light eight times round the world while a person could count " one."

547. *At what rate of velocity does the light of the stars travel?*

At the same velocity as all other light. And yet there are stars so distant that, although the light of the sun reaches the earth in eight minutes and a half, it requires *hundreds of years* to bring their light to us.

548. *What is the relative intensity of primary and reflected light?*

The intensity of a reflection depends upon the power of the reflecting surface. But, taking the sun and moon as the great examples of primary and reflected light, the intensity of the *sun's light* is 801,072 times *greater than that of the moon.*

549. *What is polarized light?*

Polarized light is light which has been subjected to *compound refraction*, and which, after polarization, exhibits a new series of phenomena, differing materially from those that pertain to the primary conditions of light.

550. *What are the chief deductions from the phenomena observed under the polarization of light?*

The polarization of light appears to confirm in a high degree the vibratory *theory of light;* and to show that the vibrations of light have two planes or directions of motion. The mast of a ship, for instance, has two motions: it progresses *vertically* as the ship is impelled forward, and it rolls *laterally* through the motion of the billows.

Something like this occurs in the vibrations of light, only the *vertical vibration* is the condition of *one ray*, and the *lateral vibration* is the condition of another ray, and the vibrations of these two rays intersect each other in the solar ray. When these vibrations occur together, the ray has certain properties and powers. But by polarization the rays may be *separated*, and the result is two distinct rays, having *different vibrations.*

It then appears that various bodies are transparent to these polarized rays *only in certain directions.* And this fact is supposed to show that bodies are made up of their atoms arranged in certain planes, through or between which the *lateral* or the *vertical* waves of light, together or singly, can or cannot pass; and that the transparency or the opacity of a body is determined by the *relation of its atomic planes* to *the planes of the vibrations of light.*

Ordinary light, passing through transparent media, produces no very remarkable effect in its course; but *polarized light* appears to illuminate every atom of the permeated substance, and by surrounding it with a prismatic clothing, to afford an illustration of its *molecular arrangement.*

551. *Why are two persons able to see each other?*

Because rays of light *flow from their bodies to each other's eyes,* and convey an impression of their respective conditions.

In some popular works that have come under our notice, we find that the student is told that "we cannot absolutely see each other—we only *see the rays of light reflected from each other.*" The statement is erroneous as expressed. We do not see the *rays* of light, for if we did so, the effect of vision would be destroyed, and all bodies would *appear* to be in a state of *incandesence,* or of *phosphoresence.* Rays of light, which are in themselves *invisible,* radiate from the objects we look upon, enter the pupil of the eye, and impress the seat of vision in a manner which conveys to the mind a knowledge of the form, colour, and relative size and position of the figure we look upon. If this is not seeing the object—*what is?* It would be just as reasonable to say, that we cannot *hear* a person speak—that we only hear the *vibrations of the air.* But as the vibrations are imparted to the air by the organs of voice of the speaker, as he sets the air in motion, and makes the air his messenger to us, we certainly hear *him,* and can dispense with any logical myths that confound the understanding, and contribute to no good result.

552. *What is actinism?*

Actinism is the chemical property of light.

Actinism—ray power.

553. *Why does silver tarnish when exposed to light?*

Because of the *actinic,* or chemical power of the rays of the sun.

554. *Why do some colours fade, and others darken, when exposed to the sun?*

Because of the *chemical* power of the sun's rays.

555. *Why can pictures be taken by the sun's rays?*

Because of the actinic powers that accompany the solar light.

556. *What is the particular chemical effect of light exhibited in the production of photographic pictures?*

Simply the *darkening of preparations of silver, by the actinic rays.*

557. *Why are photographic studios usually glazed with blue glass?*

"The hay appeareth, and the tender grass showeth itself, and herbs of the mountain are gathered."—PROV. XXVII.

Because blue glass obstructs many of the luminous rays, but it is perfectly transparent to *actinism*.

558. *Why do plants become scorched under the unclouded sun?*

Because the heat rays are in excess. The clouds shut off the scorching light; but, like the blue glass of the photographer's studio, they transmit *actinism*.

559. *What effect has actinism upon vegetation?*

It quickens the germination of seeds; and assists in the formation of the colouring matter of leaves. Seeds and cuttings, which are required to germinate quickly, will do so under the effect of blue glass (which is equivalent to saying, the effect of an increased proportion of *actinism*), in half the time they would otherwise require.

560. *In what season of the year is the actinic power of light the greatest?*

In the *spring*, when the germination of plants demands its vitalising aid. In *summer*, when the maturing process advances, *light* and *heat* increase, and *actinism* relatively declines. In the *autumn*, when the ripening period *arrives*, *light* and *actinism* give way to a greater ratio of *heat*.

We shall have frequently, in the progress of our lessons, to refer to *light* in its connection with the chemistry of nature, and with organic life. But let us now invite the student to pause, and for a moment contemplate the wonders of a sunbeam. How great is its velocity—how vast its power—how varied its parts—yet how ethereal! First, let us contemplate it as a simple beam in which *light* and *heat* are associated. How deep the darkness of the night, and how that darkness clings to the recesses of the earth. But the day beams, and darkness flies before it, until every atom that meets the face of day is lit up with radiance. That which before lay buried in the shade of night is itself now a radiator of the luminous fluid. Mark the genial warmth that comes as the sister of light; then stand by the side of the experimentalist and watch the point on which he directs the shining focus, and in an instant see iron melt and stones run like water, under the fervent heat! Now look upward to the heavens, where the falling drops of rain have formed a natural prism in the rainbow, and shown that the beam of pure whiteness, refracted into various rays, glows with all the tints that adorn the garden of nature. These are the visible effects of light. But follow it into the crust of the earth, where it is, by another power, which is neither light nor heat, quickening the seed into life watch it as the germ springs up, and the plant puts forth its tende

"But as it is written, Eye hath not seen, nor ear heard, neither have entered into the heart of man, the things which God hath prepared for them that love him."—CORINTH. BOOK I., II.

parts, touching them from day to day with deeper dyes, until the floral picture is complete. Follow it unto the sea, where it gives prismatic tints to the *anemone*, and imparts the richest colours to the various *algae*. Think of the millions of pictures that it paints daily upon the eyes of living things. Contemplate the people of a vast city when, attracted by some floating toy in the air, a million eyes look up to watch its progress. The sun paints a million images of the same object, and each observer has a perfect picture. It makes common to all mankind the beauties of nature, and paints as richly for the peasant as for the king. The Siamese twins were united by a living cord which joined their systems, and gave unity and sympathy to their sensations. In the great flood of light that daily bathes the world, we have a bond of union, giving the like pleasures and inspirations to millions of people at the same instant. And that which floods the world with beauty, should no less be a bond of unity and love.

CHAPTER XXVII.

561. *What is electricity?*

Electricity is a property of *force* which resides in all matter, and which constantly seeks to establish an *equilibrium.*

562. *Why is it called electricity?*

Because it first revealed itself to human observation through a substance called, in the Greek language, *electrum.* This substance is known to us as *amber.*

563. *In what way did electrum induce attention to this property of force in matter?*

Thales, a Greek philosopher, observed that, by briskly rubbing *electrum,* it acquired the property of *attracting* light particles of matter, which moved towards the amber, and attached themselves to its surface, evidently under the influence of a *force* excited in the amber.

564. *What is amber?*

It is a *resinous* substance, hard, bitter, tasteless, and glossy. It has been variously supposed to be a vegetable gum, a fossile, and an animal product. It is probably formed by a *species of ant* that inhabit pine forests. The bodies of ants are frequently found in its substance

"He made darkness his secret place: his pavilion round about him were dark waters and thick clouds of the skies.

565. *Why does the rubbing of a stick of sealing-wax cause it to attract small particles of matter?*

Because it excites in the sealing-wax that *force* which was first observed in the *amber*. Sealing-wax, therefore, is called an *electric* (*amber-like*) body.

566. *Why do we hear of the electric fluid?*

Simply because the term *fluid* is the most convenient that can be found to express our ideas when speaking of the *phenomena of electric force*. But of the nature of electricity, except through its observed *effects*, nothing is known.

567. *What substances are electric?*

All substances in nature, from the *metals* to the *gases*. But they differ very widely in their electrical qualities.

568. *What is positive electricity?*

Electricity, when it exists, or is excited, in any body, to an amount which is *in excess* of the amount natural to that body, is called *positive* (called also *vitreous*).

569. *What is negative electricity?*

Electricity, when it exists, or is excited, in any body, in an amount which *is less* than is the amount natural to that body, is called *negative* (called also *resinous*).

570. *Why is "positive" electricity called also "vitreous." and "negative" electricity called also "resinous"?*

Because some philosophers believe that there is but *one electricity*, but that it is liable to variations of *quantity* or *state*, which they distinguish by *positive* and *negative;* while other philosophers believe that there are *two electricities*, which they name *vitreous* and *resinous*, because they may be induced respectively from *vitreous* and *resinous* substances, and they display forces of attraction and repulsion.

571. *Upon what do the electrical phenomena of nature depend?*

Upon the tendency of *electricity* to find an *equilibrium* between its *positive* and *negative* states (assuming there to be but *one* fluid):

or upon the tendency of *vitreous electricity* to seek out and combine with *resinous electricity* (assuming that there are *two* fluids).

572. *How does the equilibrium of electricity become disturbed?*

By changes in the condition of matter. As electricity resides in all substances, and is, perhaps, an essential ingredient in their condition, so every change in the state of matter—whether from heat to cold, or from cold to heat; from a state of rest to that of motion; from the solid to the liquid, or the aeriform condition, or *vice versa;* or whether substances combine chemically and produce new compounds—in every change *the electrical equilibrium is disturbed;* and, in proportion to the degree of disturbance, is the force exerted by electricity to resume its balance in the scale of nature.

573. *How does electricity seek to regain equilibrium?*

By passing through substances that are favourable to its diffusion; therefore they are called *conducting* or *non-conducting* bodies, according as they favour or oppose the transmission of the electrical current.

574. *What substances are conductors of electricity?*

Metals, charcoal, animal fluids, water, vegetable bodies, animal bodies, flame, smoke, vapour, &c.

575. *What substances are non-conductors?*

Rust, oils, phosphorous, lime, chalk, caoutchouc, gutta percha, camphor, marble, porcelain, dry gases and air, feathers, hair, wool, silk, glass, transparent stones, vitrefactions, wax, amber, &c. These bodies are also called *insulators.* Some of these substances, as chalk, feathers, hair, wool, silk, &c., though non-conductors when *dry*, become conductors when *wetted.*

Insulating—preventing from escaping.

576. *Why are amber and wax classed among the non-conductors, when they have been pointed out as electrics, and used to illustrate electrical force?*

It is *because* they are *non-conductors* that they have displayed, under excitement, the attractive force shown in respect to the particles of matter which were drawn towards their substances. If a bar of *iron* were excited, instead of a stick of wax, electricity would be equally developed; but the iron, *being a good conductor*, would pass the electricity to the hand of the operator as fast as it accumulated, and the equilibrium would be undisturbed.

577. *What is the effect when electricity, in considerable force, seeks its equilibrium, but meets with insulating bodies?*

The result is a violent action in which, *intense heat and light* are developed, and in the evolution of which *the electric force becomes expended.*

578. *What is the cause of electric sparks?*

The electric force, passing through a conducting body to find its *equilibrium*, is checked in its course by an insulator, and emits a spark.

579. *What produces the electric light?*

Currents of electricity pass towards each other along wires at the ends of which two charcoal points are placed. As long as the charcoal points remain in contact, the electric communication is complete, and no light is emitted, but, when they are drawn apart, intense heat and light are evolved.

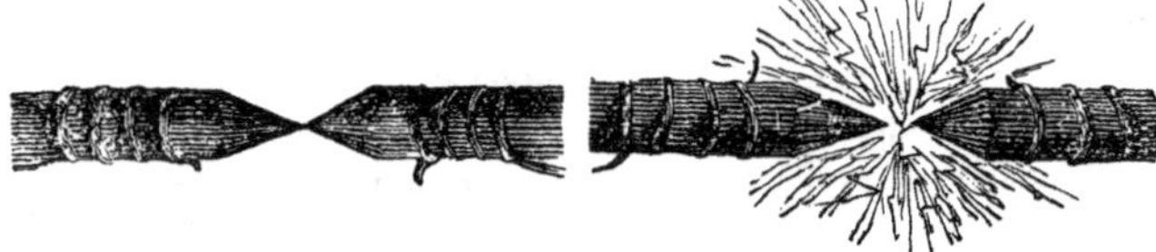

Figs. 17 & 18.—SHOWING THE EFFECT OF THE UNION AND THE SEPARATION OF THE CHARCOAL POINTS.

580. *What is the cause of lightning?*

Lightning is the result of *electrical discharges* from the *clouds*.

581. *What developes electricity in the clouds?*

Evaporations from the surface of the earth; changes of temperature in the atmospheric vapour; chemical action upon the earth's surface; and the friction of volumes of air of different densities against each other.

582. *Why do these phenomena produce electricity?*

Because they disturb the equilibrium of the electric force, and produce *positive* and *negative* states of electricity.

583. *When does lightning occur?*

When clouds, charged with the *opposite electricities* approach, the forces rush to each other, and combine in a state of equilibrium.

584. *Why does lightning attend this movement of the forces of electricity?*

Because the atmosphere, being unable to convey the great charges of electricity as they rush towards each other, *acts as an insulator*, and lightning is caused by the *violence of the electricity in forcing its passage.*

585. *Does lightning ever occur when the conducting power is equal to the force of the electricity?*

No; electricity passes invisibly, noiselessly, and harmlessly, whenever it finds a sufficient source of *conduction.*

CHAPTER XXVIII.

586. *Why does lightning sometimes travel through a "zig-zag" course?*

Because the electricity, being resisted in its progress by the air, *flies from side to side*, to find the readiest passage.

587. *Why does lightning sometimes appear forked?*

Because, being resisted in its progress by the air, the electricity divides into two or more points, and *seeks a passage in different directions.*

588. *Why is lightning sometimes like a lurid sheet?*

Because the flash is distant, and therefore we see only the *reflection.*

"He directeth it under the whole heavens, and his lightning unto the ends of the earth.

589. *When is the flash of lightning straight?*

When the distance between the clouds whose electricities are meeting, is small.

590. *What is the cause of the aurora borealis?*

The ***mingling of the electricities*** of the higher regions of the atmosphere.

591. *When does the flash of lightning appear blue?*

When the degree of electrical excitement is intense, and ***general throughout the atmosphere.***

592. *Why does lightning sometimes appear red, at others yellow, and at others white?*

Because of the varying humidity, which affects the ***refracting power*** of the atmosphere.

593. *Does lightning ever pass upwards from the earth to the clouds?*

Yes; when the earth is charged with a ***different electricity*** to that which is in the clouds.

594. *Does lightning ever pass directly from the clouds to the earth?*

Yes; when the electricity of the clouds seeks to combine with the ***different electricity*** of the earth.

The mingling of the electricities of the earth and the air must be continually going on. But *lightning* does not attend the phenomena, because all natural bodies, vapours, trees, animals, mountains, houses, rocks, &c., &c., act more or less as conductors between the earth and the air. It is only when there is a great disturbance of the electrical forces, that ***terrestrial lightning*** is developed. When lightning strikes the earth with great force, it sometimes produces what are called ***fulgurites*** in sandy soils; these are hollow tubes, produced by the melting of the soil.

595. *What is the extent of mechanical force of lightning?*

Lightning has been proved, in one instance, to have struck a church with a force equal to more than 12,000 horse-power. A single horse-power, in mechanical calculations, is equivalent to raising a weight of 32,000 lbs. one foot in a minute. The force of lightning, therefore, has been proved to be equal to the raising of 381,000,000 lbs. one foot in a minute. This is equal to the united power of twelve of our largest steamers, having collectively 24 engines of 500 horse-power each. The velocity of electricity is so great that it would travel round the world *eight times in a minute.*

The church alluded to was St. George's church, Leicester, a new edifice, which was completely destroyed on the 1st of August, 1846, by a thunderstorm. The steeple was rent asunder, and massive stones were hurled to a distance of thirty feet. The vane rod and top part of the spire fell down perpendicularly and carried with it all the floors of the tower. A similar disaster occurred to St. Bride's church, Fleet-street, London, about 100 years ago. The lightning first struck upon the metal vane of the steeple, and then ran down the rod and attacked the iron cramps, smashing the large stones that lay between them. The church was nearly destroyed. By the same wonderful force, ships have been disabled, trees split asunder, houses thrown down, and animals struck dead.

596. *Why is it dangerous to stand near a tree during an electric storm?*

Because the tree is a *better conductor than air*, and electricity would probably strike the tree, and then pass to the person standing near.

597. *If trees are good conductors, why do they not convey the electricity to the ground?*

Trees are only *indifferent conductors*, and the electricity would quit the tree to pass through any *better conductor.*

598. *Why is it dangerous to sit near a fire during an electric storm?*

Because the chimney, being *a tall object*, and smoke a *good conductor*, would probably attract the electricity, and convey it to the body of a person sitting near the fire.

599. *Why is it dangerous to be near water during an electric storm?*

Because water is a *good conductor*, and the vapour arising from it might attract the electricity. Man, *being elevated over the water*, might form the first point attacked by the electricity.

600. *Are iron houses dangerous during an electric storm?*

No; they are *very safe*, because their entire surface is a good conductor, and would convey the electricity harmlessly to the earth.

601. *Why does electricity seize upon bell wires and iron fastenings?*

Because copper wires are the *very best* conductors of electricity; and iron articles are also good conductors.

602. *Supposing electricity to attack a bell wire, where would the point of danger exist?*

At the *extremities of the wire,* where the conducting power of the wire would cease, and the electricity would seek to find *another conductor.*

603. *Are umbrellas, with steel frames, dangerous in an electric storm?*

They are dangerous *in some degree,* because they might convey electricity to the hand, and then transfer it to the body. But, generally speaking, when it rains, the rain itself, *being a good conductor,* relieves the disturbance of electricity by conveying it to the ground.

604. *Are iron bedsteads dangerous in electric storms?*

No, they are safe, because the iron frame, completely surrounding the body, and having *a great capacity for conduction,* would keep the electricity away from the body.

605. *Why is it safe to be in bed during an electric storm?*

Because *feathers, hair, wool, cotton, &c.,* especially when dry, are good *insulators* or *non-conductors.*

606. *What is the safest situation to be in during an electric storm?*

In the centre of a room, *isolated* as far as possible from surrounding objects; sitting on a chair, and avoiding handling any of the conducting substances. The windows and doors should be closed, to prevent *drafts of air.*

607. *In the open air, what is the safest situation?*

"God thundereth marvellously with his voice: great things doeth he, which we cannot comprehend."—JOB XXXVI.

To keep aloof, as far as possible, from elevated structures; and regard the rain, though it might saturate our clothes, as a protection against the lightning stroke, for *wet clothes* would supply so *good a conductor*, that a large amount of electricity would pass over man's body, through wet garments, and he would be quite unconscious of it.

During a violent electric storm in the Shetland Islands, a fishing boat was attacked by the electric fluid, which tore the mast to shivers. A fisherman was sitting by the side of the mast at the time, but he felt no shock. Upon taking out his watch, however, he found that the electric current had actually fused his watch into a mass. In this case, it is more than probable that the man was saved through the saturation of his clothes with rain.

608. *Do lightning conductors "attract" electricity?*

Not unless the electric current lies in their vicinity.

609. *Why have lightning conductors sometimes been found ineffective?*

Because they have been unskilfully constructed; have been too small in their dimensions, and have not been properly laid to convey the electricity harmlessly away.

610. *What is the best metal for a lightning conductor?*

Copper, the conducting power of which is *five times greater than that of iron.*

611. *Why should a large building have several conductors?*

Because the influence of a conductor over the electricity of the surrounding air does not extend to more than a radius of double the height of the conductor above the building: for instance, a conductor rising *ten feet high* above the building would influence the electricity *twenty feet all round the conductor.*

612. *Why should conductors have at their base several branches penetrating the earth?*

To facilitate the discharge of the *accumulated electricity* into the earth.

613. *Why does electricity affect the shapes of clouds?*

Because electricity does not penetrate the *masses* of *bodies,* but affects generally *their surfaces.* Hence electricity exists in the

"All ye inhabitants of the world, and dwellers on the earth, see ye, when he lifteth up an ensign on the mountains; and when he bloweth a trumpet, hear ye."—ISAIAH XVIII.

surfaces of clouds, and in its efforts to *find an equilibrium* it causes the clouds to roll in *heavy masses*, having dark outlines.

The fact that electricity resides in, and is conducted by, the *surfaces* of bodies, is well established, and should receive due attention in the protective measures adopted to secure life and property against the effects of lightning. A practical suggestion that arises out of this fact is, that *tubes of copper* would form far more efficient conductors than *bars* of the same metal. A *copper tube*, of half an inch diameter, would conduct *nearly double* the amount of electricity which could be conveyed away by a *bar* of copper of the same diameter. The upper extremity of the tube should be open obliquely, that the electric current might be induced to pass over *both the inner and outer surfaces*.

CHAPTER XXIX.

614. *What is thunder?*

Thunder is the *noise which succeeds the rush* of the electrical fluid through the air.

615. *Why does noise follow the commotion caused by electricity?*

Because, by the violence of the electric force, vast *fields of air are divided;* great volumes of air are *rarefied;* and *vapours* are *condensed*, and thrown down as *rain*. Thunder is therefore caused by the *vibrations of the air*, as it collapses, and seeks to restore its own *equilibrium*.

616. *Why is the thunder-peal sometimes loud and continuous?*

Because the electrical discharge take place near the hearer, and therefore the vibrations of the air are heard in their full power.

617. *Why is the thunder-peal sometimes broken and unequal?*

Because the electrical discharge takes place at a *considerable distance*, and the vibrations are affected in their course by *mountains* and *valleys*. Because, also, the *forked arms* of the lightning strike out, in different directions, causing the sounds of thunder to reach us from *varying distances.*

618. *Why has the thunder-peal sometimes a low grumbling noise?*

Because the electrical discharges, though violent, take place *far away*, and the vibrations of the air *become subdued.*

619. *Why does the thunder-peal sometimes follow immediately after the flash of lightning?*

Because the discharge of electricity takes place near the hearer.

620. *Why does the thunder-peal sometimes occur several seconds after the flash?*

Because the discharge takes place far away, and *light* travels with a much greater velocity than *sound.*

621. *Through what distance will the sound of thunder travel?*

Some *twenty or thirty miles*, according to the *direction of the wind*, and the violence of the peal.

622. *Through what distance will the light of lightning travel?*

The *light* of lightning, and its reflections, will penetrate through a distance of from *a hundred and fifty to two hundred miles.*

623. *How may we calculate the distance at which the electric discharge takes place?*

Sound travels at the rate of *a quarter of a mile in a second.* If therefore, the peal of thunder is heard *four seconds* after the flash of lightning, the discharge took place about a mile off. The pulse of an adult person beats about *once in a second;* therefore, guided by the pulse, any person may calculate the probable distance of the storm:—

2 beats, ½ a mile.
3 beats, ¾ of a mile.
4 beats, 1 mile.

"The clouds poured out water; the skies sent out a sound; thine arrows also went abroad.

5 beats, $1\frac{1}{4}$ miles.
6 beats, $1\frac{1}{2}$ miles.
7 beats, $1\frac{3}{4}$ miles.
8 beats, 2 miles, &c.

Attention should be paid to the *direction and speed of the wind*, and some modifications of the calculation be made accordingly. Persons between 20 and 40 years of age should count *five beats of the pulse to a mile;* under 20, *six beats.*

624. *Why are electric storms more frequent in hot than in cold weather?*

Because of the *greater evaporation,* as the effect of heat; and also of the *effect of heat* upon the particles of all bodies.

625. *Why do electric storms frequently occur after a duration of dry weather?*

Because *dry air,* being a bad conductor, prevents the *opposite electricities* from finding *their equilibrium.*

626. *Why is a flash of lightning generally succeeded by heavy rain?*

Because the electrical discharge destroys the *vescicles* of the vapours. If a number of *small soap-bubbles* floating in the air were *suddenly broken* by a violent commotion of the atmosphere, the *thin films* of the bubbles would form *drops of water,* and fall *like rain.*

627. *Why is an electrical discharge usually followed by a gust of wind?*

Because the equilibrium of the atmosphere is disturbed by the *heat and velocity of lightning,* and the *condensation of vapour.* Air, therefore, rushes towards those parts where a degree of *vacuity* or *rarefaction* has been produced.

628. *What is a thunderbolt?*

The name *thunderbolt* is applied to an electrical discharge, when the lightning appears to be developed with the greatest intensity around a nucleus, or centre, as though it contained a burning body. But there is, in reality, *no such thing as a thunderbolt*

"The voice of the Lord is upon the waters: the God of glory thundereth; the Lord is upon many waters."—PSALM XXIX.

629. *Why do electric storms purify the air?*

Because they restore the *equilibrium of electricity* which is essential to the salubrity of the atmosphere; they intermix the *gases of the atmosphere*, by agitation; they *precipitate the vapours* of the atmosphere, and with the precipitation of vapours, *noxious exhalations* are taken to the earth, where they become absorbed; they also contribute largely to the formation of *ozone*, which imparts to the air corrective and restorative properties.

630. *What is ozone?*

Ozone is an *atmospheric element* recently discovered, and respecting which differences of opinion prevail. It is generally supposed to be *oxygen* in a state of *great strength*, constituting a variety of form or condition.

631. *Why do we know that electricity contributes to the formation of ozone?*

Because careful observations have established the fact that the proportion of *ozone* in the atmosphere is *relative to the amount of electricity*.

632. *What are the properties of ozone?*

It displays an extraordinary power in the neutralisation of putrefactions, rapidly and thoroughly counteracting noxious exhalations; it is the most powerful of all *disinfectants*.

Schonbien, the discoverer of *ozone*, inclines to the opinion that it is a *new chemical element*. Whatever it may be, there can be no doubt that it plays an important part in the economy of nature. Its *absence* has been marked by pestilential ravages, as in the *cholera* visitations; and to its *excess* are attributed epidemics, such as *influenza*. It was found, during the last visitation of cholera, that the *fumigation of houses with sulphur* had a remarkable efficacy in preventing the spread of the contagion. The combustion of sulphur ozonised the atmosphere; the same result occurs through the emission of *phosphoric vapours;* ozone is also developed by the electricity evolved by the *electrical machine*, and in the greater *electrical phenomena of nature*. The smell imparted to the air during an electric storm is identical with that which occurs in the vicinity of an electrical apparatus—it is a *fresh* and *sulphurous* odour. The opinion is gaining ground that the respiration of animals and the combustion of matter are sources of ozone, and that plants produce it when under the influence of the direct rays of the sun. It is also believed to be produced by water, when the sun's rays fall upon it. The most recent opinion respecting ozone is, that it is *electrized oxygen*. The subject is of vast importance, and opens another field of discovery to the pioneers of scientific truth.

"The voice of thy thunder was in the heaven: the lightnings lightened the world, the earth trembled and shook."—PSALM LXVII.

633. *What is magnetism?*

Magnetism is *the electricity of the earth*, and is characterised by the circulation of *currents of electricity passing through the earth's surface.*

634. *What are magnetic bodies?*

Magnetic bodies are those that exhibit phenomena which show that they are under the influence of *terrestrial electricity*, and which indicate the direction of the *poles*, or *extreme points, of magnetic force.*

635. *What is Galvanism?*

Galvanism is the action of *electricity upon animal bodies*, and is so called from the name of its first discoverer, Galvani.

636. *What is Voltaic electricity?*

Voltaic electricity is the electricity that is developed during *chemical changes*, and is so called after Volta, who enlarged upon the theory of Galvani.

637. *What are the differences between mechanical, or frictional electricity, Voltaic electricity, Galvanism, and magnetism?*

Frictional electricity is electricity *suddenly* liberated under the effects of the *motion*, or the mechanical disturbance of bodies.

Voltaic electricity is a *steady flow* of an electric current, arising from the *gradual changes* of *chemical* phenomena.

Galvanism and *Voltaism* are almost *identical*, since the latter is founded upon, and is a development of, the former. But the term *Galvanism* is frequently used when speaking of the development of electricity in *animal bodies.*

Magnetism is the electricity of the *earth*, and is understood to imply the *fixed electricity of terrestial bodies.*

Man knows not *what electricity is;* yet, by an attentive observance of its *effects*, he avails himself of the power existing in an unknown source, and produces marvellous results. When the Grecian philosopher, Thales, sat rubbing a piece of amber, and watching the attraction of small particles of matter to its surface, he little knew of the mighty power that was then whispering to him its offer to serve mankind. And when Franklin, with the aid of a boy's plaything, drew down an electric current from the clouds, and caught a spark upon the knuckles of his hand, *even he* little conjectured that the time was so near when that strange element, which sent its messenger to him along the string of a kite, would become one of man's most submissive servants.

So many great results have sprung from the careful observation of the simplest phenomena, that we should never pass over inattentively the most trifling thing that offers itself to our examination. Nature, in her revelations, never seeks to *startle* mankind. The formation of a rock, and the elaboration of a truth, are alike the work of ages. It was the simple blackening of silver by the sun's rays which led to the discovery of the *chemical agency of light*. It was the falling of an apple which pointed Newton to the discovery of the *laws of gravitation*. It was the force of steam, observed as it issued from beneath the lid of a kettle, that led to the invention of the *steam-engine*. And it is said of Jacquard, that he *invented the loom* which so materially aided the commerce of nations, while watching the motions of his *wife's fingers*, as she plied her knitting. As great discoveries spring from such small beginnings, who among us may not be the herald of some great truth—the founder of some world-wide benefaction?

That the area of discovery has not perceptibly narrowed its limits, is evident from the fact that the greatest elements in nature are still mysteries to man. And though it may not be within the power of a finite being to unravel the chain of wonders that enfold the works of an infinite God,—still it is evident, from the progress which discovery has made, and from the good which discovery has done, that God *does* invite and encourage the human mind to contemplate the workings of Divine power, and to pursue its manifestations in every element, and in every direction.

The wonderful force of *electricity* astonishes us all the more when we view it in contrast with that equally wonderful element, *light*. We have seen that light travels with a velocity of 192,000 miles in a second, but that it falls upon a delicate balance so gently, that it produces no perceptible effect. As far as we know the nature of *electricity*, it is even *more ethereal* than *light;* yet, while the *ether of light* falls harmlessly and imperceptibly—even with the momentum of a flight of *ninety-five millions of miles*, the *ether of electricity*, bursting from a cloud only *five hundred yards* distant, will split massive stones, level tall towers with the dust, strike majestic trees to the ground, and instantly extinguish the life of man! *Why* does *the one ether* come divested of all mechanical force, while that which seems to be *even more ethereal* than it, is capable of exerting the mightiest force over material things? Does it not appear that the Creator of the universe has established these paradoxes of power to testify his Omnipotence—to show to man that with Him all things are possible; and that, in the grand cosmicism of the universe, every attribute of Omnipotence has been fulfilled?

Let us now consider man's relation to this Omnipotence. He sees that electricity smites the tall edifice, and observes that in doing so it displays a choice of a certain substance through which it passes harmlessly, and that its violence is manifested only when its path is interrupted. Man, taking advantage of this preference of electricity for a particular conductor, stretches out an arm of that substance, and points it upwards to the clouds; electricity accepts the invitation, and passes harmlessly to the earth. But this not all: man learns by observation that electricity resides in all matter; that it may be collected or dispersed; that it travels along a good conductor at the rate of *half-a-million of miles in a second of time;* he constructs a battery, a kind of scientific fortress, in which he encamps the great warrior of nature; and then, laying down a conducting wire, he liberates the mighty force: but its flight must be on the path which man has defined, and its journey must cease at the terminus which man has decreed, where, by a simple contrivance of his ingenuity (the movements of a magnetic needle), the electric current is made to deliver whatever message of importance he desires to convey. Thus, the element which in an instant might deprive man of life, is subdued by him, and made the obedient messenger of his will

CHAPTER XXX.

638. *What is the atmosphere?*

The *atmosphere* is the transparent and elastic body of mixed gases and vapours which envelopes our globe, and which derives its name from Greek words, signifying *sphere* of *vapour.*

639. *To what height does the atmosphere extend?*

It is estimated to extend to from *forty to fifty miles* above the surface of the earth.

640. *Why is it supposed that the atmosphere does not extend beyond that height?*

Because it is found, by experiment and observation, that the air becomes *less dense* in proportion to its altitude from the earth's surface. The gradual decrease of atmospheric density observed in ascending a mountain, or in a balloon, supplies sufficient data to enable us to calculate the height at which the atmosphere would probably *altogether cease.*

At an altitude of 18,000 feet the air is indicated by the barometer to be only *half as dense* as at the surface of the earth. And as the densities of the atmosphere decrease in a geometrical progression, the density will be reduced to *one-fourth* at the height of 36,000 feet; and to *one-eighth* at 54,000 feet. The effects of the decreasing density of the atmosphere are, that the *intensity of light and sound are diminished, and the temperature is lowered.* Persons who have reached a very high elevation, state that the sky above them began to assume the appearance of darkness; and there can be no doubt that, if it were possible to reach an altitude of some fifty to sixty miles, there would be *perfect black-*

"For he looketh to the ends of the earth, and seeth under the whole heaven,
To make the weight for the winds."—JOB XXVIII.

ness, although the sun's rays might be pouring through the darkened space, to illuminate the atmosphere. Upon the summit of Mont Blanc, the report of a pistol at a short distance can *scarcely be heard*. When Gay Lussac reached the height of 23,000 feet, he breathed with great pain and difficulty, and felt distressing sensations in his ears, as though they were *about to burst*. Upon the high table-lands of Peru, the lips of Dr. Ischudi cracked and burst; and blood flowed from his eyelids.

641. *What is the amount of atmospheric pressure at the earth's surface?*

The pressure of the atmosphere at the earth's surface is *fifteen pounds* to every square inch of surface. That is to say, that the column of air, extending fifty miles over a square inch of the earth, presses upon that square inch with a weight equal to *fifteen pounds.*

642. *Is that the weight of dry or moist air?*

That is the weight of air at what is called the *point of saturation*, when it is fully charged with *watery vapour*.

643. *What is the proportion of watery vapour in the atmosphere?*

The proportion *constantly varies*. Evaporation is not a result of accident; it seems an *established law* that the air shall constantly *absorb vapour* until it has reached the maximum that it can hold. Experiments have been tried, in which dry air has been pressed upon the surface of water with great force, *but no degree of pressure could prevent the formation of vapour*. (*See* 431.)

644. *What is the total amount of atmospheric pressure on the earth's surface?*

The total amount of atmospheric pressure on the earth's surface, at 15lbs. to the square inch, amounts to 12,042,604,800,000,000,000lbs. This pressure is equal to that of a globe of lead of *sixty miles in diameter*.

645. *What is the pressure of the atmosphere upon the human body?*

Estimating the surface of man's body to be equal to *fifteen square feet*, he sustains an atmospheric pressure of 32,400lbs., or nearly *fourteen tons and a-half*. The mere *variation of weight*, arising out of the changes in the state of the atmosphere, may amount to as much as a *ton and a-half.*

646. *Why does not man feel this pressure?*

Because the diffusion of air which, *surrounding him in every direction,* and acting upon the *internal* as well as the *external* surfaces of his body, and probably *surrounding every atom of his frame,* establishes an equilibrium, in which every degree of pressure *counteracts and sustains itself.*

647. *What is the weight of air relative to that of water?*

A cubic foot of air weighs only 523 grains, a little more than *an ounce*; a cubic foot of water weighs *one thousand ounces.*

648. *What is the greatest height in the atmosphere which any human being has ever reached?*

M. Gay Lussac, in the year 1804, ascended to the height of 23,000 feet.

649. *What is a vacuum?*

A vacuum is a space *devoid of matter.* The term is generally applied to those instances in which air is drawn from within an air-tight vessel.

650. *Is it possible to form a perfect vacuum?*

It is probably *impossible to do so,* even with the most powerful instruments—some portion of air would remain, but in so thin a a form that it would be *imperceptible.*

651. *Why does the depression of a pump-handle cause the water to flow?*

Because the putting down of the handle lifts up the piston with its *valve closed,* thereby tending to produce a *vacuum;* but *the pressure of the air* upon the water *not contained in the pump,* forces more water up into the part where a *vacuum* would otherwise be formed. Then, when the handle is raised, and the piston forced downwards, *the valve opens,* and the water rushes through.

There is a second valve, below the piston, which closes with the downward movement, to prevent the water from *rushing back again.*

652. *How high will atmospheric pressure raise water in the bore of a pump?*

It will raise water to an elevation of *thirty feet* above its level.

653. *Why will it raise water to an elevation of thirty feet?*

Because a column of water of *thirty feet high*, nearly balances the weight of *a column of air* of equal surface, *extending to the whole height of the atmosphere.* When, therefore, water is elevated to the height of thirty feet, the power of the pump is enfeebled, as the air and the water *balance each other.*

654. *How is water raised to a greater elevation when it is required?*

By mechanical contrivances, by which the water is *forced* to a greater elevation.

655. *Why does water run through the bent tube called a syphon?*

Because the atmospheric pressure upon the water on *the outside of the syphon* forces it into the tube as fast as the syphon empties itself through its longer arm.

656. *Why does water run through the longer arm of the syphon?*

Because the weight of the water in the longer arm of the syphon *is greater than that in the shorter;* therefore it runs out by its own gravity. And, as in running out, it creates a tendency towards a *vacuum,* the pressure of the outer air comes into operation, and forces the water through the tube.

657. *Why does water issue from the earth in springs?*

Some springs are caused by *natural syphons* formed in the fissures of rocks, which, communicating with bodies of water, are continually filled by atmospheric pressure, and therefore convey streams of water to the point where they are set free.

"Ascribe ye strength unto God: his excellency is over Israel, and his strength is in the clouds."—PSALM LVIII.

658. *Why, if a wine glass is filled with water, and a card laid upon it, and the whole inverted, will the water remain in the glass?*

Because the pressure of the atmosphere upon the surface of the card counteracts the weight of the water.

659. *What has the card to do with the experiment?*

It forms *a base* upon which the water may rest, while the glass is being inverted; and it prevents the air from acting upon the *fluidity* of the water, and forcing it out of the glass.

660. *Why will not beer run out of the tap of a cask until a spile has been driven in at the top?*

Because the pressure of the air upon the opening of the tap counteracts the weight of the beer. But when the spile is driven in, the air enters at the top, *and counteracts its own pressure at the bottom.*

661. *Why does a cup in a pie become filled with juice?*

Because *the heat expands the air*, and drives nearly all of it out of the cup. When the pie is taken out of the oven, and begins to cool, air cannot get into the cup again, because its edges are surrounded by juice. A *partial vacuum*, therefore, exists within the cup, and the pressure of the external air *forces the juice into it.*

662. *Does the cup prevent the juice from boiling over?*

No. So long as the *heat* exists, the cup remains *empty*; and as it occupies space, the air is driven out of it, into the pie, it rather tends to force the juice over the sides of the dish. It is only *when cooling* that the juice enters the cup.

663. *Why can flies walk on the ceiling?*

Because their feet are so formed that they can form a *vacuum*, under them; their bodies are therefore sustained in opposition to gravitation by *atmospheric pressure.*

664. *How did Mr. Sands perform the feat of walking across the ceiling?*

By having large discs of wet leather attached to his feet, so that when they were placed upon a smooth surface, the air was excluded, and when he allowed his weight to act upon one of the discs, it formed a *hollow cup* and a *vacuum*. By forming a vacuum of only *twelve square inches* he gained a pressure of 180lbs.; this being more than his weight he could accomplish the feat with no other difficulty than that of remaining in an inverted position. The air was admitted underneath the discs by valves, which were closed by springs, which being pressed by the heels of the performer, let in the air, and *set the feet free.*

665. *Why is it difficult to strike limpets from rocks?*

Because they have the means of forming a *vacuum* under their shells, and are pressed on to the rocks by the weight of the atmosphere.

666. *Why can snails move over plants in an inverted position?*

Because they form a *vacuum* with the smooth and moist surfaces of their bodies, and are supported by atmospheric pressure.

CHAPTER XXXI.

666. *What is wind?*

Wind is air *in motion.* (*See* 234.)

667. *What are the velocities of winds?*

A *breeze* travels ten feet in a second; a *light gale,* sixteen feet in a second; a *stiff gale,* twenty-four feet in a second; a *violent squall,* thirty-five feet in a second; *storm wind,* from forty-three to fifty-four in a second; *hurricane* of the temperate zone, sixty feet in a second; *hurricane* of the torrid zone, one hundred and twenty to three hundred feet in a second. When wind flies at one mile an hour, it is scarcely perceptible. When its velocity is one hundred miles an hour, it tears up trees, and devastates its track.

668. *What are trade winds?*

Trade winds are vast currents of air, which *sweep round the globe* over a belt of some 12,000 miles in width.

669. *What is the cause of trade winds?*

The air over the tropical regions becomes heated and ascends; it then diverges in two high currents, one towards the north, and the other towards the south pole, where, being cooled, it again descends, and returns towards the equator to replace the air as it ascends therefrom. There is, therefore, a constant revolution of vast currents of air between the tropics and the poles, producing *north and south winds.*

670. *Why do the trade winds blow from east to west, though, in their origin, their direction is from north to south and from south to north?*

Because, as the north and south winds blow towards the equator, they are affected by the revolution of the earth from *west to east.* As the two winds from the poles approach the equator, they are gradually diverted from their northerly and southerly course, to an easterly direction, by the revolution of the earth.

671. *Why is there a prevalence of calms at the equator?*

Because, as the north and the south winds move towards the equator, they drive before them volumes of atmosphere, which, meeting in opposite directions, resist and counterpoise each other, and abide in a state of stillness between the north and south-easterly winds, one on the north and the other on the south of the equator.

672. *What are monsoons?*

Monsoons are *periodical winds* which blow at a given period of the year from one quarter of the compass, and in another period of the year from the opposite quarter of the compass.

673. *What is the cause of monsoons?*

Monsoons are caused by changes in the position of the sun. When the sun is in the southern hemisphere, it produces a *north-east wind,* and when it is in the northern hemisphere, a *north-west wind* The north-east monsoon blows from November to March.

and the south-west monsoon from the end of April to the middle of October. The region of moonsoons lies a little to the north of the northern border of the trade wind, and they blow with the greatest force, and with most regularity, between the eastern coast of Africa and Hindustan.

674. *What determines the character of winds ?*

The character of winds is influenced by the condition of *the surfaces over which they blow.* Winds blowing over dry and arid plains and deserts are *dry and hot.* Winds blowing across snow-capped mountains and regions of ice are *cold.* Winds that cross oceans are *wet ;* and those that cross extensive continents are *dry.*

675. *What winds are most prevalent in England ?*

In England out of a *thousand days,* north winds prevail in 82; north-east, 111; east, 99; south-east, 81; south, 111; south-west, 225; west, 171; north-west, 120.

676. *What is the cause of storms ?*

Storms result from violent commotions of the atmosphere, and are chiefly the result of extreme *changes of temperature.*

The *magnetic* state of the earth, and the *electrical* state of the atmosphere, also materially influence the phenomena of storms.

By some persons the theory is entertained that storms result from various winds *rushing into a centre* in which the atmosphere has become extremely condensed. According to this theory. a storm is a mighty whirlwind.

A most violent hurricane occurred in 1780, which destroyed Lord Rodney's fleet, and a vast number of merchant ships. It is said to have killed 9,000 persons in Martinique alone, and 6,000 in St. Lucia. The town of St. Pierre in Martinique was totally destroyed; and only fourteen houses in the town of Kingston, in St. Vincent, were left uninjured.

677. *Why do the most violent storms occur in and near the tropics ?*

Because there the temperature is very high, and the cold currents of air rushing towards the equator from the poles, causes great *atmospheric disturbance*

678. *What are whirlwinds ?*

"Out of the south cometh the whirlwind; and cold out of the north."—
JOB XXXVII.

Whirlwinds are produced by violent and contrary currents meeting and striking upon each other, producing *a circular motion.* They generally occur after long calms, attended by much heat.

Whirlwinds occurring at sea, or over the surface of water, sometimes put the water in motion, and as the wind rises upwards it lifts with it a whirling mass of water, producing a *water spout.*

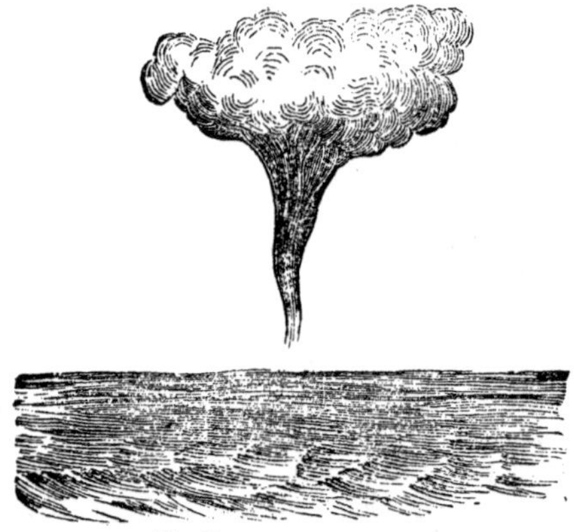

Fig. 19.—A WATER SPOUT.

679. *Why does the chimney smoke when the fire is first lighted?*

Because the air in the chimney is of the same temperature as that in the room, and therefore *will not ascend.*

680. *Why does the smoking (into the room) cease, after the fire has been lighted a little while?*

Because the air in the chimney, being warmed by the fire beneath, becomes lighter and ascends rapidly.

681. *Why does a long chimney create a greater draught than a short one?*

Because the short chimney contains *less air* than the long one, there is, consequently, less difference of weight between the warm

"And, lo, the smoke of the country went up as the smoke of a furnace."—
GEN. XIX.

air of the short chimney and the external air; it therefore has not so great an *ascensive power*.

682. *Why does smoke issue in folds and curls?*

Because it is *pressed upon* by the *cold air* which always *rushes towards a rarer atmosphere.* It thus illustrates the development of *storms.*

683. *Why do some chimneys smoke when the doors and windows are closed?*

Because the draught of air is not sufficient to supply the wants of the fire, and enable it to create an *upward current.*

684. *What is the best method of conveying air to fires?*

Tubes built in the walls, communicating with the outer air, and terminating *underneath the grates.*

685. *Why is this the best method of ventilation?*

Because doors and windows may then be made air-tight, and *draughts across rooms be prevented.*

686. *Why do chimneys that stand under elevated objects, such as hills, trees, and high buildings, smoke?*

Because the wind, striking against the elevated object, *flies back*, and a part of it *rushes downward.*

687. *Why do sooty chimneys smoke?*

Because the accumulation of the soot *diminishes the size of the flue*, and lessens the ascensivé power of the draught, by reducing the quantity of *warm air.* It also obstructs the motion of the air, by the *roughness of its surface*

688. *Why do chimneys smoke in damp and gusty weather?*

Because the ascending air is *suddenly chilled* by gusts of damp and cold air, and driven down the chimney.

"Remember that thou magnify his work, which men behold. Every man may see it; man may behold it afar off."—JOB XXXVI.

689. *Why does smoke ascend in a straight line in mild and fine weather?*

Because the air is still, and being dry and warm it *does not chill the smoke*, nor drive it out of its course.

690. *Why do the wings of wind-mills turn round?*

Because the wind, striking *at an angle* upon the wings, forces them aside; and as there are four wings all upon the same angle, and fixed upon the same centre, the *oblique pressure* of the wind causes the centre to rotate.

There is a world of *miniature phenomena* which has never been fully recognised, in which we may see the mightier works of nature pleasingly and truthfully illustrated.

When the wind blows into the corner of a street, and whirling around, catches straw, dust, and feathers in its arms, and then wheels away, flinging the troubled atoms in all directions,—it is a miniature of the mightier *whirlwind*, which wrecks ships, uproots trees, and levels houses with the earth.

When a cloud of dust, on a hot summer's day, rises and flies along the thirsty road, making the passenger close his eyelids, and dusting the leaves of wayside vegetation,—it is a miniature of the terrible *simoom*, which blows from the desert sands, scattering death and devastation in its track.

When steam issues from the tea-urn, and becomes condensed in minute drops upon the window-pane,—the miniature is of the *earth's heat*, evaporating the waters, and the cold air of night condensing the vapours into *dew*.

When grass and corn bend before the wind, and are beaten down by its force; when the pond forgets its calm, and rises in troubled waves, casting the flotilla of natural boats that move upon its surface, in rude disorder upon its windward shore,—the little storm is but a miniature of those great *hurricanes* which wrecked a fleet in the Black Sea, and levelled the encampments of a mighty army.

When the snow that has gathered upon the house-top, warming beneath the smiles of the sun, slips from its bed, and drops in accumulated heaps from the roof,—it is a miniature of those terrible *avalanches* which in the Pyrenees bury villages in their icy pall, and doom man and beast to death.

When the rivulet hurries on its course, and meeting with obstructions, leaps over them in mimic wrath, overturning some little raft upon which, perchance, a weary fly has alighted,—it is a miniature of those *rapids* on whose banks the hippopotamus and the alligator yet live; and where, though rarely, man may be seen directing his raft over the troubled current, amid the rush of *debris* from forests unexplored.

And when, in a basin of the rivulet, two opposing currents meet, and form a little vortex into which insect life and vegetable fragments coming within the

"Can any understand the spreadings of the clouds, or the noise of his tabernacle?"—JOB XXXVI.

sphere of its influence are drawn,—it is a miniature of the roaring *whirlpool* or the wilder *maelstrom* of the Norwegian seas.

Nature rehearses all her parts in mild whispers; and for every picture that she paints, she places a first study upon the canvas. Man need not go into the heart of her terrors to understand their laws. Many an unknown Humboldt, sitting by the river's side, may rejoice in the "aspects of nature," and share the bliss of knowledge with the great philosopher.

CHAPTER XXXII.

691. *What is a barometer?*

A barometer is an instrument which ***indicates the pressure of the atmosphere***, and which takes its name from two Greek words signifying *measurer of weight.*

692. *Why does a barometer indicate the pressure of the atmosphere?*

Because it consists of a tube containing ***quicksilver***, closed at one end and open at the other, so that the pressure of the air upon the open end balances the weight of the column of mercury (quicksilver), and when the pressure of the air upon the open surface of the mercury increases or decreases, the mercury ***rises or falls*** in response thereto.

693. *Why is a barometer called also a "weather-glass?"*

Because changes in the weather are generally preceded by ***alterations in the atmospheric pressure.*** But we cannot perceive those changes as they gradually occur; the alteration in the height of the column of mercury, therefore, enables us to know that atmospheric changes are taking place, and, by observation, we are enabled to determine certain rules by which *the state of the weather may be foretold* with considerable probability.

694. *Why are barometers constructed with circular dials, and an index to denote changes?*

Because that is a convenient mechanical arrangement, by which

" Fair weather cometh out of the north: with God is terrible majesty."—
JOB XXXVII.

the alterations of the relative pressures of the air and the mercury are *more clearly denoted than by an inspection of the mercury itself.*

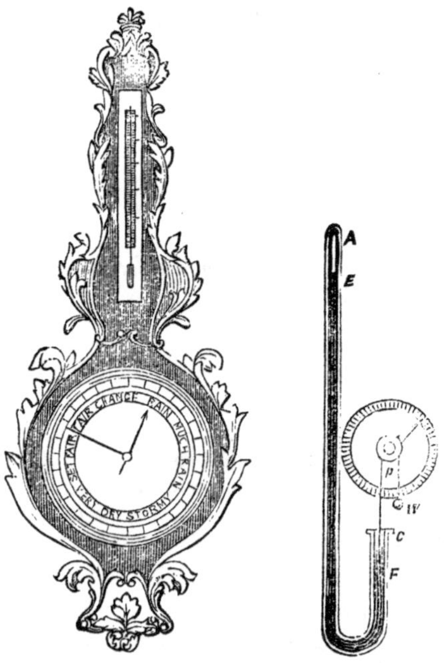

Fig. 20.—BAROMETER. Fig. 21.—TUBE OF BAROMETER, WHEEL, AND PULLEY.

605. *Why does the hand of the weather dial change its position when the column of mercury rises or falls?*

Because a weight, which *floats upon the open surface of the mercury*, is attached to a string, having a nearly equal weight at the other extremity; the string is laid over a revolving pivot to which the hand is fixed, and *the friction of the string turns the hand, as the mercury rises or falls.*

"Thou visitest the earth, and waterest it: thou greatly enrichest it with the river of God, which is full of water: thou preparest them corn, when thou hast so provided for it."—PSALM LXV.

696. *Why does tapping the face of the barometer sometimes cause the hand to move?*

Because the weight on the surface of the mercury frequently *leans against the sides of the tube*, and does not move freely. And, also, the mercury clings to the sides of the tube by *capillary attraction;* therefore, tapping on the face of the barometer *sets the weight free*, and overcomes the attraction which *impedes the rise or fall of the mercury.*

Fig. 21 illustrates the mechanism at the back of the barometer. A is a glass tube; between A and E there exists a *vacuum*, caused by the weight of the mercury pressing downwards. This space being a vacuum, makes the barometrical column more sensitive, as there is no internal force to resist or modify the effects of the external pressure. E represents the height of the column of mercury; C the open end of the tube; F the weight resting on the surface of the mercury; P the pivot over which the string passes, and upon which the hand turns: W the weight which forms the pulley with the weight F.

697. *Which is the heavier, dry or vaporised air?*

Dry air is *heavier* than air impregnated with vapours.

698. *Why is dry air heavier than moist air?*

Because of the *extreme tenuity of watery vapours,* the density of which is *less than that of atmospheric air.*

699. *Why does the fall of the barometer denote the approach of rain?*

Because it shows that as the air *cannot support the full weight of the column of mercury,* the atmosphere must be thin with watery vapours.

The fall of the mercury in the long arm of the tube would cause the weight F to be pressed upwards. This would release the string to which the weight W is attached; it would, therefore, fall, and turn the hand down to Rain, or Much Rain.

700. *Why does the rise of the barometer denote the approach of fine weather?*

Because the external air becoming dense, and free from highly elastic vapours, presses with increased force upon the mercury upon which the weight F floats; that weight, therefore, sinks in the short tube as the mercury rises in the long one, and in sinking turns the hand to Change, Fair, &c.

"He caused an east wind to blow in the heaven; and by his power he brought in the south wind."—PSALM LXXVIII.

701. *Why does the barometer enable us to calculate the height of mountains?*

Because, as the barometer is carried up a mountain, *there is a less depth of atmosphere above to press upon the mercury;* it therefore falls, and by comparing various observations, it has been found practicable to *calculate the height of mountains by the fall of the mercury in a barometer.*

702. *To what extent of variation is the weight of the atmosphere liable?*

It may vary as much as *a pound and a half to the square inch* at the level of the sea.

703. *When does the barometer stand highest?*

When there is a *duration of frost,* or when *north-easterly winds* prevail.

704. *Why does the barometer stand highest at these times?*

Because the atmosphere is exceedingly *dry and dense,* and fully balances the *weight of the column of mercury.*

705. *When does the barometer stand lowest?*

When *a thaw follows a long frost;* or when *south-west winds* prevail.

706. *Why does the barometer stand lowest at those times?*

Because *much moisture exists in the air,* by which it is rendered less dense and heavy.

707. *What effect has heat upon the barometer?*

It causes the mercury to fall, *by evaporating moisture into the air.*

708. *What effect has cold upon the barometer?*

It causes the mercury to rise, by *checking evaporation,* and *increasing the density of the air.*

In noting barometrical indications, more attention should be paid to the *tendency* of the mercury at the time of the observation, than to the *actual state of the column,* whether it stands *high* or *low.* The following rules of barometic reading are given as generally accurate, but liable to exceptions:—

"For so the Lord said unto me, I will take my rest, and I will consider in my dwelling place like a clear heat upon herbs, and like a cloud of dew in the heat of harvest."—ISAIAH XVIII.

Fair weather indicated by the *rise* of the mercury.
Foul weather by the *fall* of the mercury.
Thunder, indicated by the *fall* of the mercury in *sultry weather*.
Cold, indicated by the *rise* of the mercury in spring, autumn, and winter.
Heat, by the *fall* of the mercury in summer and autumn.
Frost, indicated by the *rise* of the mercury in winter.
Thaw, by the *fall* of the mercury during a frost.
Continued bad weather, when the *fall* of the mercury has been *gradual* through several fine days.
Continued fine weather, when the *rise* of the mercury has been *gradual* through several foul days.
Bad weather of short duration, when it sets in quickly.
Fine weather of short duration, when it sets in quickly.
Changeable weather, when an *extreme* change has *suddenly* set in.
Wind, indicated by a rapid *rise* or *fall unattended by a change of temperature*.
The mercury *rising*, and the air becoming *cooler*, promises *fine weather*; but the mercury *rising*, and the air becoming *warmer*, the weather will *be changeable*.
If the top of the column of mercury appears *convex*, or curved upwards, it is an additional proof that the mercury is *rising*. Expect *fine* weather.
If the top of the column is *concave*, or curved downwards, it is an additional proof that the mercury is *falling*. Expect *bad* weather.

CHAPTER XXXIII

709. *What is the thermometer?*

The thermometer is an instrument in which *mercury* is employed to indicate *degrees of heat*. Its name is derived from two Greek words, meaning *heat measurer*.

710. *Why does mercury indicate degrees of heat?*

Because it *expands* readily with *heat*, and *contracts* with *cold*; and as it passes freely through small tubes, it is the most convenient medium for indicating *changes of temperature*.

711. *Why are there Reaumur's Thermometers and Fahrenheit's Thermometers?*

Because their inventors, after whom they are named, adopted a different system of *notation*, or *thermometrical marks*; and as their thermometers have been adopted by various countries and authors, it is now difficult to dispense with either of them.

"When ye see a cloud rise out of the west straightway ye say, There cometh a shower ; and so it is. And when ye see the south wind blow, ye say there will be heat; and it cometh to pass."—LUKE XIII.

THERMOMETER.

Reaumur. *Fahrenheit.*

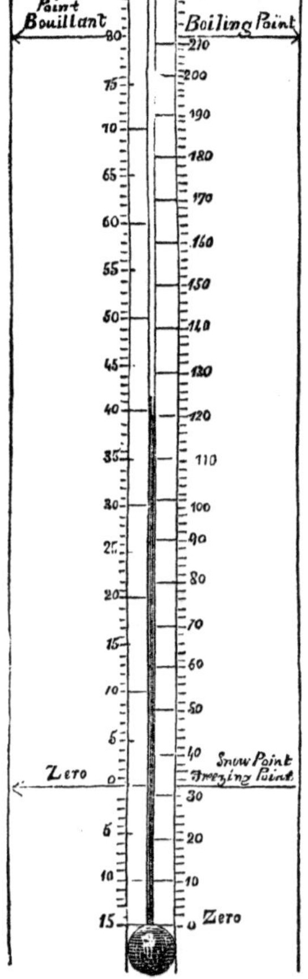

Fig. 22.—THE THERMOMETERS OF REAUMUR AND FAHRENHEIT COMPARED.

We have combined the two (see Fig. 22.) The diagram will, we have no doubt, prove exceedingly useful to scientific readers and experimentalists. There is also another system of notation, adopted by the French, called the *centigrade*, but it is not much referred to in Great Britain. In the centigrade thermometer 0 zero is the freezing point, and 100 the boiling point. Fahrenheit's scale is generally preferred. Reaumur's is mostly used in Germany. Of Fahrenheit's scale 32 is the freezing point, 55 is moderate heat, 76 summer heat in Great Britain, 98 is blood heat, and 212 is the boiling point. Mr. Wedgwood has invented a thermometer for testing *high temperatures*, each degree of which answers to 130 degrees of Fahrenheit. According to his scale cast iron melts at 2,786 deg.; fine gold at 2,016 deg.; fine silver 1,873 deg.; brass melts at 1,869 deg.; red heat is visible by day at 980 deg.; lead melts 612 deg.; bismuth melts 476 deg.: tin melts 442 deg.; and there is a curious fact with regard to the three metals, lead, bismuth, and tin, that if they are mixed in the proportions of 5, 8, and 3 parts respectively, the mixture (after previous fusion) will melt at a heat below that of boiling water.

712. *What is the difference between the thermometer and the barometer?*

In the thermometer the column of mercury is much smaller than in the barometer, and is sealed from the air; while in the barometer the column of mercury is open at one end to atmospheric influence.

713. *Why does the mercury in the thermometer, being sealed up, indicate the external temperature?*

Because the heat passes through the glass, in which the mercury is enclosed, and *expanding or contracting the metal within the bulb*, causes the small column above it to *rise* or *fall.*

714. *When does the thermometer vary most in its indication of natural temperature?*

It varies more in the *winter* than in the *summer* season.

715. *Why does it vary more in the winter than in the summer?*

Because the temperature of our climate *differs more from the temperature of the torrid zones in the winter* than it does in the *summer*, and the *inequalities of temperature* cause frequent changes in the degree of prevailing heat.

The same remarks (714, 715,) apply to the barometer.

CHAPTER XXXIV.

716. *What is sound?*

Sound is an *impression produced upon the ear* by *vibrations* of *the air.*

717. *What causes the air to vibrate and produce sounds?*

The atoms of *elastic bodies* being caused to *vibrate* by the application of some kind of force, *the vibrations of those atoms are imparted to the air*, and sound is produced.

718. *How do we know that sounds are produced by the vibrations of the air, induced by the vibrations of the atoms of bodies?*

If we take a tuning fork, and hold it to the ear, we hear *no sound.* If we move it rapidly through the air, or if we blow upon it, it produces *no sound*; but if we *strike it, a sound immediately occurs*; the vibration of the fork may be seen, and felt by the hand that holds it; and *as those vibrations cease*, the sound *dies away.*

719. *How do we know that without air there would be no sound?*

"And even things without life giving sound, whether pipe or harp, except they give a distinction in the sounds, how shall it be known what is piped or harped."—CORINTH, XIV.

Because if a tuning fork were to be struck in a *vacuum* (as under the receiver of an air pump) *no sound* would be heard, although the *vibrations* of the fork could be *distinctly seen.*

720. *How are the vibrations of sonorous bodies imparted to the air?*

When a bell is struck, the force of the blow gives an instant agitation to all its particles. The air around the bell is driven back by the impulse of the force, and thus a *vibration of compression* is imparted to the air; but the air returns to the bell, by its own natural elasticity, thus producing a *vibration of expansion*—when it is again struck, and thus *successive vibrations* of compression and expansion are *transmitted through the air.*

721. *How rapidly are these vibrations transmitted through the air?*

They travel at a rate of rather more than *a quarter of a mile in a second,* or *twelve miles and three-fourths in a minute.*

722. *Do all sounds travel at the same rate?*

All sounds, whether strong or weak, high or low, musical or discordant, *travel with the same velocity.*

723. *Why are bells and glasses stopped from ringing by touching them with the finger?*

Because the contact of the finger *stops the vibration* of the atoms of the metal and glass, which therefore *cease to impart vibrations to the air.*

724. *Why does a cracked bell give discordant sounds?*

Because the *connection* between the atoms of the bell being *broken,* their vibrations are not uniform: some of the atoms vibrate *more intensely* than the others; the vibrations imparted to the air are therefore *jarring* and *discordant.*

725. *Why, when we see a gun fired at a distance, do we see the flash and smoke, before we hear the report?*

Because *light,* which enables us to *see,* travels at the velocity of

"My heart maketh a noise in me: I cannot hold my peace, because thou hast heard, O my soul, the sound of the trumpet, the alarm of war."—JER. IV.

192,000 miles in a *second;* while *sound*, by which we *hear*, travels only at the rate of a quarter of a mile in a *second*.

726. *Why does the tread of soldiers, when marching in long ranks, appear to be irregular?*

Because the sounds proceeding from *different distances*, reach our ears in *varying periods of time*.

727. *What are the numbers of vibrations in a second that produce the various musical sounds?*

C or Do, 480 vibrations in a second; B or Si, 450 vibrations; A or La, 400 vibrations; G or Sol, 360 vibrations; F or Fa, 320 vibrations; E or Mi, 300 vibrations; D or Re, 270 vibrations; C or Do, 240 vibrations. It is thus seen that the *more rapid* the vibrations, the *higher* the note, and *vice versa*.

728. *Why does the length of a wire or string determine the sound that it produces?*

Because the *shorter the string* the *more rapid* are its vibrations when struck.

729. *Why does the tension of a wire or string affect its vibrations?*

Because when the string or wire is tight, a touch communicates vibrations to *all its particles*; but when it is loose the vibrations are *imperfectly communicated*.

730. *Why are some notes low and solemn, and others high and quick?*

Because the vibrations of musical strings vary from 32 vibrations in a second, which produces a soft and deep bass, to 15,000 vibrations in a second, which produces the sharpest treble note.

731. *Why can our voices be heard at a greater distance when we speak through tubes?*

Because the vibrations are *confined to the air within the tube*, and are not interfered with by *other vibrations* or movements in the air; the tube itself is also a *good conductor of sound*.

"And I will cause the noise of thy songs to cease; and the sound of thy harps shall no more be heard."—EZEKIEL XXVI.

732. *Is air a good conductor of sound?*

Air is a *good conductor*, but water is a *better conductor than air;* wood, metals, the earth, &c., are also *good conductors.*

733. *Why can we hear sounds at a greater distance on water than on land?*

For various reasons: because the smooth surface of water is a good conductor; because there are fewer noises, or counter vibrations, to interfere with the transmission of sound; and because there are no elevated objects to impede the progress of the vibrations.

734. *Why do sea-shells give a murmuring noise when held to the ear?*

Because what may be called *expended vibrations* always exist in air where various sounds are occurring. These *tremblings* of the air are received upon the thin covering of the shell, and thus being *collected into a focus*, are *transmitted to the ear.*

735. *Why can people in the arctic regions converse when more than a mile apart?*

Because there the air, being *cold and dense*, is a very good conductor; and the *smooth surface of the ice* also favours the *transmission of sound.*

736. *Why do savages lay their heads upon the earth to hear the sounds of wild beasts, &c.?*

Because the earth is a good conductor of sound. For this reason, also, persons *working under ground in mines*, can hear each other digging at considerable distances.

737. *Why can church clocks be heard striking much more clearly at some times than at others?*

Because the density of dry air improves the *sound-conducting power* of the atmosphere. The transmission of sounds is also assisted by the direction of the winds.

738. *Why may the scratching of a pin at one extremity of*

a long pole be heard by applying the ear to the opposite extremity?

Because wood is a good conductor of sound, and its atoms are *susceptible of considerable vibration.* It is, therefore, chosen in numerous instances for the construction of *musical instruments.*

Deaf persons have been known to derive pleasure from music by placing their hands upon the wood-work of musical instruments while being played upon.

739. *Why is the hearing of deaf persons assisted by ear-trumpets?*

Because ear-trumpets *collect the vibrations of the air* into a focus, and make the sounds produced thereby more intense.

740. *Why are sounding-boards used to improve the hearing of congregations?*

Because, being suspended over, and a little behind, the speaker, they *collect the vibrations* of the air, and *reflect* them towards the congregation.

741. *What are echoes?*

Echoes are sounds *reflected* by the objects on which they strike.

742. *Why do some echoes occur immediately after a sound?*

Because the reflecting surface is *very near;* therefore the sound returns immediately.

743. *Why do some echoes occur a considerable time after a sound?*

Because they are at a considerable distance, and the sound takes time to travel to it, and an equal time to return.

744. *Why do some echoes change the tone and quality of sound?*

Because the reflecting surface, having vibratory qualities of its own, *mingles its own vibrations with that of the sound.*

745. *Why are there sometimes several echoes to one sound?*

"And God said, Let the waters under the heaven be gathered together unto one place, and let the dry land appear: and it was so."—GEN. I.

Because there are various *reflecting surfaces*, at different distances, each of which returns an echo.

746. *Are sounds reflected only by distant objects?*

Sounds are doubtless reflected by *walls and ceilings* around us. But we do not perceive the echoes, because they are so near that they occur at the same moment with the sound. In lofty buildings, however, there is frequently a *double sound*, making the utterance of a speaker indistinct. This arises from the echo following very closely upon the sound.

747. *Why, when we are walking under an arch-way or a tunnel, do our voices appear louder?*

Because the sounds of our voices are *immediately reflected*. And as a *gas reflector increases the intensity of light*, so *a sound reflector* will *increase the apparent strength of our voices*.

There are many places where remarkable echoes occur. On the banks of the Rhine, at Lurley, if the weather be favourable, the report of a rifle, or the sound of a trumpet, will be repeated at different periods, and with various degrees of strength, from crag to crag, on opposite sides of the river alternately. A similar effect is heard in the neighbourhood of some of the Lochs in Scotland. There is a place at Woodstock, in Gloucestershire, which is said to echo a sound fifty times. Near Rosneath, a few miles from Glasgow, there is a spot where, if a person plays a bar of music upon a bugle, the notes will be repeated by an echo, but a third lower; after a short pause, another echo is heard, again in a lower tone; then follows another pause, and a third repetition follows in a still lower key. The effect is very enchanting. The whispering galleries of St. Paul's, of the cathedral church of Gloucester, and of the Observatory of Paris, owe their curious effects to those laws of the reflection of sound, by which echoes are produced; but in these cases the effect is assisted by the elliptical form of the edifice, each person being in the focus of an ellipse.

CHAPTER XXXV.

748. *What is water?*

Water is a fluid composed of *two* volumes of *hydrogen* to *one* of *oxygen*, or *eight* parts by weight of *oxygen* to *one* of *hydrogen*. It is nearly colourless and transparent.

749. *Why, if a saucer of water be exposed to the air, will it gradually disappear?*

hold there ariseth a little cloud from the sea, of the bigness of a man's hand. And it came to pass in the meantime, that the heaven was black with clouds and wind, and there was a great rain."—1 KINGS XVIII.

Because water is highly expansive, and *rises in thin vapour,* when in contact with warm and dry air.

750. *Why does steam issue from the spout of a kettle?*

Because the heat of the fire passes into the water, and *drives* its atoms apart, making those of them that rise quickly to the surface *lighter than the air,* upon which they consequently rise.

751. *Why does water become solid when it freezes?*

Because the *latent heat of the water* passes away from between its atoms into the air; the atoms, therefore, draw closer together.

752. *Why, if the atoms of water draw closer together when freezing, does ice expand, and occupy greater space than water?*

Because, *when the atoms of water are congealing,* they do not form a *compact mass,* but arrange themselves *in groups of crystal points,* which occupy greater space. Water *contracts* when freezing until it sinks to 40 deg., and then it *expands* as ice is formed.

32 deg. is said to be the *freezing* point, but it should be called the *frozen* point.

753. *Why does water boil?*

Because heat, *entering into the lower portions* of the water, *expands* it; the heated portions are then *specifically lighter* than those that are cooler; the hot water therefore *rises upward, and forces the cooler water down.*

754. *What proportion of the earth's surface is covered with water?*

There are about one hundred and forty seven millions of square miles of *water,* to forty-nine and a half millions of square miles of *land.*

755. *What is the amount of water pressure?*

The pressure of the sea, at the depth of 1,100 yards, is equal to 15,000 lbs. *to the square inch.*

"But the land, whither ye go to possess it, is a land of hills and valleys, and drinketh water of the rain of heaven."—DEUT. XI.

756. *What element is the most abundant in nature?*

Oxygen, which forms so large a part of *water*. Of animal substances, *oxygen* forms *three-fourths;* of *vegetable substances* it forms *four-fifths;* of *mineral substances* it forms *one-half;* it forms *eight-ninths* of the *waters* and *one-fifth* of the *atmosphere;* and aggregating the whole creation, from *one half to two-thirds* consists of *oxygen.*

757. *In what ways does man use oxygen?*

Man *eats, drinks, breathes,* and *burns* it, in various proportions and combinations. It is estimated that *the human race* consume in those various ways 1,000,000,000lbs. daily; that the *lower animals* consume double that amount; and that, in the varied works of nature, no less than 8,000,000,000lbs. of *oxygen* are used *daily.*

758. *Why does water dissolve various substances?*

Because the *atoms of water* are very minute; they therefore *permeate the pores,* or spaces, between the atoms of those bodies, and *overcoming their attraction for each other,* cause them to separate.

759. *Why does hot water dissolve substances more readily than cold?*

Because the *heat assists to repel the particles* of the substance undergoing solution, and *gives the water a freer passage* between the atoms.

760. *Why is pump water sometimes hard?*

Because, in passing through the earth, it has become impregnated with mineral matters, usually the *sulphate* and *carbonate of lime.*

761. *Why is rain water soft?*

Because it is derived from vapours which, in ascending to the clouds, *could not bear up the mineral waters with them.* It therefore became purified or distilled.

762. *Why do kettles become encrusted with stony deposits?*

Because that portion of the water which is driven off in steam *leaves the mineral matters behind;* they therefore form a crust around the sides of the kettle.

It is said that if a child's marble be placed in a kettle, it will attract the earthy particles, and prevent the encrusting of the sides of the vessel.

763. *Why is it difficult to wash in hard water?*

Because the soap unites with the mineral matters in the water, and being *neutralised* thereby, cannot *dissolve the dirt* which we desire to cleanse away.

764. *Why is the sea salt?*

Because salt is a mineral which prevails largely in the earth, and which, *being very soluble in water,* is taken up by the ocean.

Lakes and rivers, also, even those that are considered fresh, hold in solution *some degree of saline matters,* which they contribute to the ocean.

As, in the evaporations from the sea, the salt remains in it, while the vapours fall as rain, and again wash the earth and carry some of its mineral properties to the ocean, the *greater saltness of the sea,* as compared with rivers, is accounted for.

By some persons the opinion is entertained that the sea has been *gradually getting salter* ever since the creation of the world. This, they say, arises from the evaporation of water free from salt, and the returns of the water to the sea, taking with it salt from the land.

765. *What is the estimated amount of salt in the sea?*

The amount of common salt in the various oceans is estimated at 3,051,342 *cubic geographical miles,* or about five times more than the mass of the mountains of the Alps.

766. *What is the depth of the sea?*

The extreme depth has not, probably, been ascertained. But Sir James Ross took soundings about 900 miles west of St. Helena, whence he found the sea to be nearly *six miles in depth.* Now, if we take the height of the highest mountain to be five miles, the distance from that extreme rise of the earth, to the known depth of the sea, will be no less than *eleven miles.*

767. *Why are the waters of some springs impregnated with mineral matters?*

Because the water passes through beds of soda, lime, magnesia, carbonic acid, oxides of iron, sulphate of iron, &c., &c., and *takes up in some slight degree the particles of those minerals,* according to the proportions in which they abound.

768. *Why does iron rust rapidly when wetted?*

Because the water contains a large proportion of oxygen, some of which combines with the iron and forms *an oxide of iron,* which is *rust.*

769. *Why does stagnant water become putrid?*

Because the *large amount of oxygen* which it contains accelerates the decomposition of dead *animal and vegetable substances* that accumulate in it.

770. *Is there danger in drinking water on account of the living animalcules which it contains?*

No danger arises from the *living creatures* in water; but *putrefactive* matters may produce serious diseases.

771. *What is the best method of guarding against impurities?*

By obtaining water from the purest sources, and by filtering it before drinking, by which nearly all extraneous matters would be *separated from it.*

CHAPTER XXXVI.

772. *What is attraction?*

Attraction is the tendency of bodies to *draw near to each other.* It is called *attraction,* from two Latin words signifying *drawing towards.*

773. *How many kinds of attraction are there?*

There are five principal kinds of *attraction*:—

1. The attraction of *gravitation.*
2. The attraction of *cohesion.*
3. The attraction of *chemical affinity.*

"Behold, the nations are as a drop of a bucket, and are counted as the small dust of the balance: behold, he taketh up the isles as a very little thing."—ISAIAH XL.

4. The attraction of *electricity.*
5. And *capillary attraction.*

774. *Why do all bodies heavier than the air fall to the earth?*

Because they are influenced by the *attraction of gravitation,* by which all bodies are drawn towards *the centre of the earth.*

775. *Why do bodies lighter than the air ascend?*

Because the air, being a denser body, *obeys the law of attraction* and in doing so *displaces lighter bodies* that interfere with its gravitation.

776. *Why do fragments of tea, and bubbles floating upon the surface of tea, draw towards each other, and attach themselves to the sides of the cup.*

Because they are influenced by the *attraction of cohesion.*

Cohesion.—The act of sticking together.

777. *Why will a drop of water upon the blade of a knife leave a dark spot?*

Because the *iron of the knife attracts the oxygen of the water,* by *chemical affinity*; and the two substances form a thin coating of *oxide of iron.*

Affinity.—Attraction between dissimilar particles through which they form new compounds.

778. *Why do clouds sometimes move towards each other from opposite directions? and*

779. *Why do light particles of matter attach themselves to sealing wax, excited by friction?*

Because they are moved by the *attraction of electricity.*

780. *Why will a towel, the corner of which is dipped in water, become wet far above the water?*

Because the water is conveyed up through the towel, by *capillary attraction.* The atoms of the water are attracted by the *threads of the towel,* and drawn up into the *small spaces between the threads.*

Capillary.—Resembling a hair, small in diameter.

"He stretcheth out the north over the empty place, and hangeth the earth upon nothing."—JOB XXVI.

781. *Why do small bodies floating upon water move towards larger ones?*

Because the attractive power of *a large body* is greater than that of *a small one.* As each atom of matter has inherent power of attraction, it follows that a *large aggregation of particles* must attract in proportion to the number of those particles.

782. *Why do clouds gather around mountain tops?*

Because they are *attracted by the mountains.*

783. *Why would a piece of lead tied to a string, and let down from a church steeple, incline a little from the perpendicular towards the church?*

Because the *masses of stone* of which the church is built would *attract the lead.*

784. *How can man weigh the earth?*

By observing what is called the *deflection* of small bodies *when brought within given distances of larger bodies,* the degree of attraction *exercised by the large body upon the smaller one* becomes known. This attraction of the *large body* exercised over the *smaller body* is an opposing influence, *acting against the earth's attraction of the small body,* which is drawn out of its course: it constitutes a *natural balance between the influence of the earth and another body, acting in opposition to it.* Founded upon these, and some other data, man can weigh the earth, and give a morally certain result!

Deflection.—The act of turning aside.

785. *How can man weigh the planets?*

The planets exercise as certain an influence upon each other *as do two pieces of wood floating upon a basin of water.* As the planetary bodies fly through their prescribed orbits, *and approach nearer to, or travel further from, each other,* they are observed to *deviate* from that course which they must have pursued *but for the increase or the decrease of some influence of attraction.* By making observations *at various times,* and by comparing *a number of results,* it is possible *to weigh any planetary body, however vast, or however distant.*

"Is not God in the height of the heaven? and behold the height of the stars, how high they are?"—JOB XII.

786. *How can man measure the distances of the planets?*

By making observations at *different seasons of the year*, when the earth is in *opposite positions in her orbit;* and by recording, by *instruments constructed with the greatest nicety*, the *angle of sight*, at which the planetary body is viewed; by noticing, also, *the various eclipses*, and estimating *how long the first light after an eclipse has ceased* reaches the earth, it is possible to estimate the *distances* of heavenly bodies, *no matter how far in the depths of the universe those orbs may be.*

787. *What are the opinions founded upon estimates respecting the magnitude of the sun?*

The *diameter* of the *sun* is 770,800 geographical miles, or 112 times greater than the diameter of the earth; its *volume* is 1,407,124 times that of the earth, and 600 times greater than *all the planets together;* its *mass* is 359,551 times greater than the earth; and 738 times greater than that of *all the planets.* A *single spot* seen upon its surface has been estimated to extend over 77,000 miles in diameter, and a *cluster of spots* have been estimated to include an area of 3,780,000 miles.

788. *What is the weight of the earth?*

The earth has a *circumference* of 25,000 miles, and is estimated to *weigh* 1,256,195,670,000,000,000,000,000 tons.

789. *What is the specific gravity of a body?*

It is its weight estimated *relatively to the weights of other bodies.*

790. *What determines the force with which bodies fall to the earth?*

Generally speaking, their *specific gravity*, which is proportionate to the density, or *compactness of the atoms* of which they are composed.

791. *Why does a feather fall to the earth more gradually than a shilling?*

Because the *specific gravity* of the feather and of the shilling is *relative to that of the air*, the medium through which the feather and the shilling pass. If there were *no air*, a shilling and a feather dropped at the same time from a height of forty miles, *would reach the earth at the same moment.*

CHAPTER XXXVII.

792. *What is repulsion?*

Repulsion is that property in matter by which it *repels* or *recedes from*, those bodies for which it has *no attraction or affinity.*

793. *Why does dew form into round drops upon the leaves of plants?*

Because it *repels the air*, and the *substances of the leaves* upon which it rests. Because, also, its own particles *cohere.*

794. *Why do drops of water roll over dusty surfaces?*

Because they *repel* the particles of dust; and also because their own particles have *a stronger attraction for each other* than for the particles of dust.

795. *Why does a needle float when carefully laid upon the surface of water?*

Because the needle and the water *mutually repel each other.*

796. *Why does water, when dropped upon hot iron, move about in agitated globules?*

Because the *caloric* repels the particles of the water.

797. *Why does oil float upon the surface of water?*

Because, besides being specially lighter than water, the particles of the oil and the water *mutually repel each other.*

798. *What is carbonic acid?*

Carbonic acid is a mixture of *carbon* and *oxygen*, in the proportion of 3 lbs. of carbon to 8 lbs. of oxygen.

799. *Where does carbonic acid chiefly exist?*

It exists in various natural bodies in which carbon and oxygen are combined; it is evolved by the decomposition of numerous bodies called carbonates, in which carbon is united with a particular base, such as the carbonate of lime, the carbonate of iron, the carbonate of copper, &c. It is also evolved by the processes of *fermentation*, by the ***breathing of animals***, the ***combustion of fuel***, and the ***functions of plants***. Carbonic acid also ***exists in various waters***.

Carbonic acid is *found most largely in solid combinations with other bodies:* it forms 44-100ths of all limestones and marbles, and it exists in smaller quantity, combined with other earths, and with metallic oxides.

800. *What are the states in which pure carbonic acid exists?*

Pure carbonic acid may exist in the *solid*, the *liquid*, or the *aeriform* state. In the *solid state* it is produced only by artificial means, and it is then a white crystalised body, in appearance *like snow;* in the *liquid state* it is a *heavy colourless fluid;* in the *aeriform state* it is a ***pungent, heavy, colourless gas***, and is known as ***carbonic acid gas***.

801. *Why does bottled porter produce large volumes of froth, much more than the bottle could contain?*

Because, by the fermentive process, ***carbonic acid*** has been developed in the porter, and is held in *liquid solution;* but it always has a ***strong tendency to escape***, and directly the pressure is removed, it ***evolves into gas***, by which it occupies much greater space, and forces the porter in millions of small bubbles out of the bottle.

892. *Why does soda-water effervesce?*

Because ***carbonic acid gas*** is forced into the water ***by pressure***. Pressure ***alters the gas into a liquid***, and directly the pressure ceases, the liquid again ***evolves into gas***.

803. ***Why does spring water taste fresh and invigorating?***

Because it contains *carbonic acid*.

804. *Why does boiled water taste flat and insipid?*

Because the *carbonic acid* has been *driven off* by boiling.

805. *Why does beer which has been standing in a glass taste flat?*

Because its *carbonic acid* has escaped as *carbonic acid gas.*

806. *Why, when we look into a glass of champagne, do we see bubbles spontaneously appear at the bottom, and then rise to the top?*

Because, in the places where the bubbles are formed, the *liquid carbonic acid* is evolving into *carbonic acid gas.*

807. *Why do the bubbles arise from two or three points in columns, rapidly succeeding each other?*

Because, when the formation of gas once begins, and bubbles ascend, there is *less pressure* in the line of the *column of bubbles;* the carbonic acid, therefore, draws towards those points as the *easiest channel of escape.*

These explanations equally apply to the "working" of beer, by which yeast is produced; to the effervesence of various waters, acidulated drinks, ginger beer, &c., and also to the "sponging" of bread, &c.

808. *Why does gunpowder explode?*

Gunpowder is made of a very intimate *mechanical mixture* of *nitrate of potash, charcoal,* and *sulphur.* When these substances are heated to a certain degree, the nitrate of potash is decomposed, and its *oxygen* combines with the *charcoal* and *sulphur,* instantaneously forming *large volumes of carbonic acid gas* and *nitrogen,* which, seeking an escape, produce an explosion.

809. *Why does charcoal act as a powerful disinfectant?*

Because the *carbon* readily absorbs, and combines with *various gases,* neutralising their *offensive odours,* and destroying their *unhealthy properties.*

Let us now pause for a few moments to consider the importance of those two great divisions of nature, Air and Water, and to reflect upon the wisdom of some of those laws which are connected with the phenomena thereof, and which have not yet been sufficiently explained.

We have seen that the air is a thin elastic body surrounding the globe; that

"Thus saith the Lord, Let not the wise man glory in his wisdom, neither let the mighty man glory in his might, let not the rich man glory in his riches."—
JEREMIAH IX.

It consists of certain gases essential to the life of animals, and to the growth of plants; and that it takes part in most of those chemical changes, which mark the transformations of the inorganic creation. Whether it be the burning of a piece of wood, the evaporation of a drop of water, the breathing of an animal, the respiration of a plant, or the fermentation of bodies, the air in almost every instance gives or receives—and in most of the operations in which it engages, it does both.

But there is one point of view, which we must add to those which have already been considered: the order of nature consists of generation, life, and death. Every beat of the watch signals the birth of millions of living things, and the same beat proclaims that as many living organisms have yielded up their vital spark, and that forthwith the elements of which they are composed must be dissolved, and restored to the great laboratory of nature.

The air is the vast receptacle of those organic matters which are undergoing dissolution. The body of the shipwrecked mariner, cast upon the shore of a desolate island, blackens in the sun, and the full round form gradually dwindles to skin and bone, until at last the few atoms that remain crumble into dust, and are scattered to the wind. The same process occurs, with some modifications, whether bodies are buried in the earth, or dissolve upon its surface. The leaves of forests fall and accumulate in heaps, where they ferment and dissolve, leaving only their more earthy particles behind.

The amount of matter which day by day passes from the state of the living to that of the dead, must be enormous; but from the difficulties of acquiring data, beyond the possibility of calculation. Such statistics as we have, however, enable us to form conclusions as to the mighty agencies in which the air is constantly engaged. There are on the earth 1,000,000,000 inhabitants of whom nearly 35,000,000 die every year, 91,824 every day, 3,730 every hour, and 60 every minute. But *even the living die daily*, and undergo an invisible change of substance, as we shall hereafter explain.

The bodies of those many millions are dissolved in the air, in vapours and gases which, before the dissolution of each corporeal organism is complete, begin to live again in the various forms of vegetable and animal life.

Of the number of animals living and dying upon the face of the earth, we can form no adequate estimate. Of mammals there are about 2,000 ascertained species; of birds 8,000 species; of reptiles 2,000 species; of fishes some 8,000 or 10,000 species; of molluscs some 15,000 species; of shell fish 8,000 species; of insects 70,000 species. And, including others not specified here, the total number of *species* of animals probably amounts to no less than 250,000,—each species consisting of *many millions* of living creatures.

In the area of London alone, no less than 200,000 tons of fuel are annually cast into the air in the form of smoke. And if we take into account the vast operations of nature in evaporation, fermentation, and putrefactive decomposition, we may be enabled to form a conception of the mighty part which that *thin air*, of which we think so little, plays in the grand alchemy of nature.

In addition, also, to the facts already communicated, respecting the sound-bearing and light-refracting properties of air, it must be remarked, that but for the atmosphere, and the general refraction of light by its particles—each atom as it were catching a fairy taper, and dancing with it before our view—the condition of vision would be widely opposite to that which exists, and totally unsuited to our wants. The various objects upon which the illuminating rays of the sun fell, would be lighted up with an intense glare, but all around would be darkness, just as when a single ray of light is passed into a dark chamber, and directed upon a solitary object. The air, without becoming itself visible, *diffuses luminous rays*, in modified intensity, in every direction. If the air reflected so much light as to render *itself visible*, it would appear like the glittering surface of the water reflecting the solar rays, and we should then be unable to see the various objects which surround us.

Of the importance of Water in the scheme of creation, man generally entertains an imperfect conception. It is simply supposed to afford moisture to plants, drink to animals, and to promote salubrity by its cleansing properties. Let us, however, contemplate man as he stands before us, noble in form, erect in position, full of strength, joy, ambition. How much of that noble form is composed of water? Suppose that it could all be instantaneously withdrawn—not the oxygen and the hydrogen, which might combine to form water—but the fluid that exists in his body *as water*, unchanged—except by mechanical admixture with the secretions of the body—Why then that beautiful temple would collapse and become a mere shred, so thin, that it would seem but a shadow of the body as it existed before, and the beholder might doubt whether life ever inhabited a frame whose structure was so frail. It is said that *three-fourths* by weight of the human body consist of *water*. Thus, if man weighs 120lbs., 90lbs. consist of water, and this substracted, only 30lbs. of solid matter remain. This statement is rather under than over the fact.

The assertion is startling, but so true that it can be verified by simple experiment. A piece of lean flesh—say of beef—cut an inch thick, and placed in a slow oven, and allowed to remain until all its water was driven off in vapour, would become as thin as a wafer, and as light as a cork. With a more scientific arrangement, it would be possible to collect the water, and the weights of the condensed vapour, and of the solid residue, would together make up the weight of the beef: if the piece weighed sixteen ounces, the weight of the water would be about 14 ounces, and the *solid matter* about *two ounces*.

Water holds a similar proportion in the bodies of all animals, and of vegetables. It is evident, therefore, that it occupies a more important place in the scale of creation than is generally accorded to it by the unobservant mind. We are indebted to it for those atmospheric changes which constitute the peculiar feature of our varying climate. Rising in invisible vapours, it builds palaces of glory in the skies, and often presents to the view of man the imagery of heaven. Persons who have ascended above the altitude of the clouds, have described the scene upon looking down towards them as the most celestial that the mind can conceive.* Fields of fleecy radiance, majestically rolling like a sea of gold, occupied the whole range of vision, and seemed to embellish an eternity of space. Those golden clouds that at one time are decked in the richest splendour, and occupy the upper chambers of the Court of Nature, become grave councillors when the earth grows thirsty, and the plant droops with languor. They roll their heavy brows together, as in consultation upon some grave necessity: down come the refreshing showers, the mighty tongue of thunder rocks the air, the earth is drenched, and becomes fresh with the salubrity of her toilette; obnoxious substances, with their offensive exhalations, are swept away: living things rejoice, and beautiful flowers throw their incense of thanksgiving into the air: the broad blue heavens for a time look down and

"How mighty are his wonders! his kingdom is an everlasting kingdom, and his dominion is from generation to generation."—DANIEL IV.

smile upon the blessed work; and then the clouds again gather in a golden train, and one by one fill the high arches of the atmosphere, until the earth once more grows thirsty, and the flower supplicates for drink.

With reference to Light, its wonders, and the curious but imperfect theories respecting it, we have little to add, except with regard to its physiological action upon the eyes of man and of animals, which will be given in another place. But of its sister, Darkness—for it would not do now to call darkness the antagonist of light, since it will be seen that they work harmoniously for good—we have to say, that recent discoveries indicate that darkness is as necessary to the health of nature as light. Not only is it necessary to compose man and animals to sleep, to give rest to the over-wrought nerves of the industrious—but light is the quickening power of vegetation, and although plants grow by night, they grow, as man does, when stretched upon his bed—but some of their functions, which are actively excited in the presence of light, are at rest in darkness. Nor is this all: there is not an atom upon the face of the earth which is not affected by the rays of the sun, their light, their heat, their actinism. Colours change: some are bleached, others are darkened. All bodies are expanded. The hardest rock sustains *an effect* from the sun's rays; and an unceasing sun, shining upon the hardest granite, would in time produce such a disturbance of its atomic condition, that adamant would crumble away to dust.

The going down of the sun, therefore, marks the period when not only does the bird fly to her resting-place, and man turn to his couch; but when *every atom of a vast hemisphere* subsides into a state of quietude, and when homogeneous particles of matter return to their mutual rest.

In a few succeeding lessons, we intend to point out some of the scientific truths that are *illustrated in the use of toys.* We think we shall be able to show to our young readers, that even the hours of play may be made the periods of delightful instruction; and that there is *no* "reason why" the acquirement of knowledge should not sweetly accord with the occasional pursuit of those pastimes by which health of body and vigour of mind are induced.

But before we commence the discharge of that pleasant duty, let us say a few words respecting Carbon, that important agent in the world's history. It is, doubtless, perplexing to the minds of many persons, to understand how the *diamond* can be *pure carbon;* how *charcoal* can be *carbon a little less pure* than the diamond; and how *coal* and *sugar* can also be carbon, *less pure* than the charcoal. The statement that in the diamond carbon exists in a different atomic condition, is almost as instructive to the inquiring mind, as to say, "It is so, *because it is.*"

Diamonds are expensive things, and so difficult to experiment upon, even if they were not expensive, that the doors of inquiry seem locked. To turn diamonds into charcoal, or into carbonic acid gas, is a very costly formula of experiment. Charcoal fires, thus sustained, would soon burn a man out of his house; and soda water, impregnated with carbonic acid gas, produced from diamonds, would be a very expensive beverage. If we could only turn charcoal into diamonds, and carbonic acid gas into brilliants, that would be quite another affair. A new Eldorado would be discovered, and there would be so many experimenters that, when they all succeeded, they would find that diamonds had lost their value. However, as a fact for the encouragement of those who would like to be early in the race, we may state that the atoms of

"He delivereth and rescueth, and he worketh signs and wonders in heaven and in earth."—DANIEL VII.

charcoal which are repulsed from the charcoal points, during the electric agitation which produces the electric light, acquire a hardness and a sharpness almost equal to that of the diamond—only there is still the awkward obstacle in the way, that *they happen to be black.*

We must see, therefore, whether there is anything in nature that we can experiment upon, theoretically or practically, to give us a clearer conception of this difficult matter. There is a large *dew-drop* resting upon a luxuriant cabbage leaf—one of those great leaves that have flourished in defiance of the snail, and now spreads out like the gigantic frond of the *Victoria Regina.* That dew-drop is one of the beautiful diamonds which Nature sprinkles about on cloudless nights, as if to show the stars, in answer to their twinkling, that we have something that will glisten and twinkle too.

The dew-drop is a very good imitation of a diamond, and to the lover of God's works, quite as precious as the stone set in gold. It does not consist of carbon—it probably may have a mite of carbonic acid in its embrace—but that is not necessary to our purpose: all we want to know is, *the different atomic conditions* of which bodies are susceptible, and the very dissimilar appearances they exhibit under the variations of atomic states. It doesn't glisten so much as the diamond, *because it is round*—if we could cut it into a number of *facets,* it would refract light almost as perfectly as the diamond. It is not *solid*—but we can freeze it, and we shall at once exhibit two different atomic conditions, that will represent nearly enough the diamond, and the liquid carbonic acid. Then, if we evaporate the dew-drop, we shall produce a volume of vapour nearly *two thousand times as large as the dew-drop.* The steam will be white; but we have only to imagine it black, and then we get an analogy of the differences of the atomic conditions that prevail in *the diamond, carbonic acid,* and *charcoal, tinder, lamp-black,* or any light form of carbon. Of course we have been illustrating *atomic conditions only,* and not chemical composition.

There are a few other facts connected with carbon that merit consideration. Carbonic acid gas, *entering the lungs,* is a *deadly poison;* but *entering the stomach,* which lies close under the lungs, and is over-lapped by them, it is a *refreshing beverage.* Although charcoal, when burnt, gives off the most poisonous gas, it seems to be very jealous of other gaseous poisons; for if it be powdered, and set about in pans where there is a poisonous atmosphere, it will seize hold of poisonous gases, and, by absorbing, imprison them. Even in a drop of toast and water, the charred bread seizes hold of whatever impurities exist in the water; and water passed through beds of charcoal, becomes filtered, and made beautifully pure, being compelled to give up to the charcoal whatever is obnoxious. If a piece of meat that has already commenced putrifying, be sprinkled with charcoal, it will not only object to the meat putrifying any further, but it will *sweeten that which has already undergone putrefaction.* Although, in the form of gas, it will poison the blood, and cause speedy stupefaction and death; if it be powdered, and stitched into a piece of silk, and worn before the mouth as a respirator, it will say to all poisonous gases that come to the mouth with the air, "I have taken this post to defend the lungs, and I arrest you, on a charge of murderous intention." Such are the various facts connected with carbon; and they forcibly indicate that those who understand Nature's works are likely to receive her best protection.

"The father of the righteous shall greatly rejoice; and he that begetteth a wise child shall have joy of him."—PROVERBS XXIII.

CHAPTER XXXVIII.

810. *Why does a humming-top make a humming noise?*

Because the hollow wood of the top vibrates, and the edges of the hole in its sides *strike against the air as it spins;* the air is thereby set in vibration.

811. *Why does a peg-top hum less than a humming-top?*

Because, *being a solid body of wood,* and having no *hole in its sides,* its particles are *not so easily thrown into vibration;* consequently it does not so readily impart vibrations to the air.

812. *Why does a peg-top sometimes hum, and at other times not?*

Because, if it is spun with *great force,* and its peg is *struck sharply* against the pavement, *the wood is set in vibration,* and the surface of the top, repelling the air by its rapid motion, causes *vibratory waves.* But if it be spun with insufficient force, *the wood is not set in vibration.*

Fig. 23.—HUMMING-TOP BEFORE SPINNING.

Fig. 24.—HUMMING-TOP SPINNING.

813. *Why do we see the figures painted upon the humming-top, before it spins, but not while it is spinning?*

Because the rapid whirling of the top brings the images of its different parts so quickly in succession upon *the retina of the eye.*

'Train up a child in the way he should go; and when he is old, he will not depart from it."—PROVERBS XXII.

that they *deface each other*, and *impart an impression of coloured rings, instead of definite objects.*

814. *Why does a top stand erect when it spins, but fall when it stops?*

Because the top is under the influence of, and is balanced between *opposing forces.* The rapid rotation of the top gives to all its particles a tendency to *fly from the centre.* If the atoms of the wood were not held together by the *attraction of cohesion*, they would fly away in a circle outward from the top, *just as drops of water fly off from a mop, while it is being twirled.* If you take a spoonful of sand, salt, or dust, and drop it upon the top, it will be scattered in a circle, just as the atoms of the top would be, *if they were free to separate*, but not with the same force, because the atoms of the salt, &c., not being in an active state of rotation, would only be influenced *by momentary contact with the rotating body.* This tendency of the particles of a rotating body to fly outward from the centre, is called *the centrifugal force.*

Centrifugal.—From two Latin words meaning receding from the centre.

The other force influencing the top is *the attraction of gravitation:* the attraction which, were the top not spinning, would draw it towards the earth. The "spill" projecting from the bottom of the top *stands in the line in which the top is drawn towards the earth* and keeps it from obeying the law of gravitation. Therefore the rotatory motion given to the top, by the rapid unwinding of the string, and the tendency of its atoms to fly outward, *balance the top* upon the line in which it is drawn to the earth, and which is occupied by the spill, which prevents it falling to the ground.

815. *Why does a top first reel around upon the spill, then become upright, and "sleep," and then reel again, and fall?*

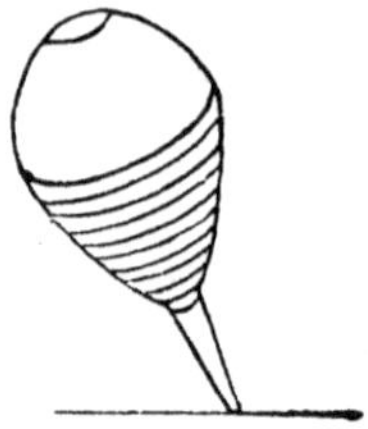

FIG. 95.—PEG-TOP "REELING."

Because, in being thrown from the hand, the top is delivered a little out of the perpendicular, but the spill *is rounded off at the point*, and when the top is rotating rapidly, the gravitative force which attracts the top to the ground continually acting upon it, *draws the weight of the top on to the extreme centre of the round point.* When the rotation subsides, and the centrifugal force is weakened, then the top *is no longer balanced upon the extreme point of the spill,* but falls upon it sides, until the force of gravitation is exerted *beyond the line of the spill,* upon the body of the top, and then it falls to the ground.

816. *Why does a top "sleep?"*

Because at that period of its spinning, which is called "sleeping," the *centrifugal* and the *gravitative forces* acting upon the top, are *nearly balanced;* and the top, obeying chiefly the *rotatory force,* appears to be in a state of comparative rest.

817. *Why does the top cease to spin?*

Because *the friction of the air against its sides,* and the *friction of the spill against the ground,* act in opposition to the *rotatory force,* which is a temporary impulse applied by external means—the hand of the person who spins it—and as soon as this *applied force* is expended, the top yields to the law of gravitation, which is *a permanent and ever-prevailing force.*

818. *Why does a marble revolve, as it is propelled along the ground?*

Because, in propelling the marble, *the thumb impels the upper surface forward, and the finger draws the under surface backward.* This gives a tendency to the upper and lower hemispheres of the marble *to separate,* which they would do, but for the *cohesion of the atoms* of the marble. The upper part of the marble, therefore, rolls forward, *drawing after it the under part,* which acquires a forward motion by the force with which it is drawn upward, and in this way the opposite portions of the marble act upon each other in the successive revolutions.

When the marble strikes upon the earth, a new influence is exerted upon it, which is *the friction of the earth* upon the surface that comes in contact with it; but the upper part of the marble, being free, *overcomes the friction acting upon the lower part,* and thus the marble continues to progress, until *the applied force which projected it is expended.*

"Better is a poor and a wise child, than an old and foolish king who will no more be admonished."—ECCLESIASTES IV.

819. *Why does a striped marble appear to have a greater number of stripes when rolling, than when at rest?*

Because the stripes are presented in *rapid succession* to the eye; and as the eye receives *fresh impressions of stripes before the previous impressions have passed away,* the stripes appear multiplied.

Fig. 26.—MARBLE AT REST

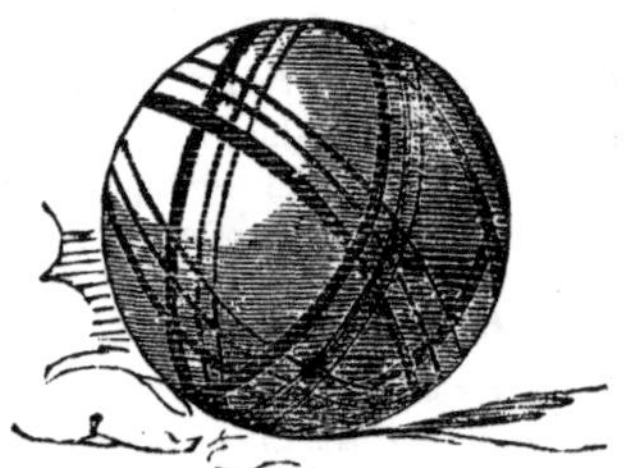

Fig. 27.—MARBLE ROLLING.

820. *Why does a marble rebound when dropped upon the pavement?*

Because the force of its fall to the earth *compresses the atoms* of which the marble is composed; and the atoms then exert the force of *elasticity to restore themselves to their former condition;* and by the exercise of this force the marble is *repelled,* or *thrown upward from the pavement.* Although a marble may be made of very hard stone, yet that stone may be *elastic,* and possess, though in a much less degree, *the same kind of elasticity which causes the India-rubber ball to rebound from the earth.*

821. *Why does a marble, assuming it to be impelled with equal force, roll further on ice than on pavement, and further on pavement than on a pebble walk?*

Because the *friction* is greater upon pavement than upon ice, and greater upon a pebble walk than upon pavement.

822. *How many forces contribute to stay the progress of a rolling marble?*

The friction of the *air*, the friction of the *earth*, and the *attraction of gravitation*, which tends to bring all bodies to a state of rest.

823. *Why do the stripes upon a marble disappear when it is spun with great velocity?*

Fig. 28.—MARBLE SPINNING RAPIDLY.

Because, as in the case of the humming-top, the different parts of the surface are *brought so rapidly in succession to the sight*, that they *deface or confuse* the impressions upon the retina.

824. *Why are rings most perceptible at the opposite points, or poles, of the marble?*

Because the point, or pole, *upon which the marble spins*, and that which *corresponds to it*, on the upper surface, travel *less rapidly* than the central portions, which being of a larger circumference, pass through a greater amount of space, in the same period of time. The stripes at the *poles* of the marble, are, therefore visible, while those at its *equator* are imperceptible. (*See* 522.)

CHAPTER XXXIX.

825. *Why are soap-bubbles round?*

Because they are *equally pressed upon all parts of their surface* by the atmosphere.

826. *Why are bubbles elongated when being blown?*

Because the *unequal pressure of the current of breath* by which they are being filled, alters the *relative pressure* upon the outer surfaces.

827. *Why does the bubble close, and become a perfect sphere, when shaken from the pipe?*

Because the *attraction of cohesion* draws the particles of soap together, directly the bubble is set free from the bowl.

Fig. 29.—BLOWING SOAP BUBBLES.

828. *Why do bubbles, blown in the sunshine, change their colours?*

Because the films of the bubbles constantly change in thickness, through the atoms from the upper part descending towards the bottom, and therefore the varying thickness of film *refracts, in different degrees, the rays of light.*

829. *Why do bubbles burst?*

Because the atoms that compose their films *fall towards the earth by gravitation;* the upper portion of the bubbles then *becomes very thin,* and as the denser air of the atmosphere *presses towards the warm breath within the bubble, it bursts the film.*

See 236. 237, *etc.*, 501, *etc.*

830. *Why do balloons ascend in air?*

Because the air or gas which they contain is *specifically lighter than the atmosphere; the atmosphere, therefore, forces itself underneath the balloon,* by its own tendency towards the earth, and the balloon is thereby raised upwards. *A balloon is but a larger kind of bubble, made of stronger materials.*

831. *Why does an air-balloon become inflated when the spirit set upon the sponge is lit?*

"A wise son heareth his father's instruction."—PROVERBS XIII.

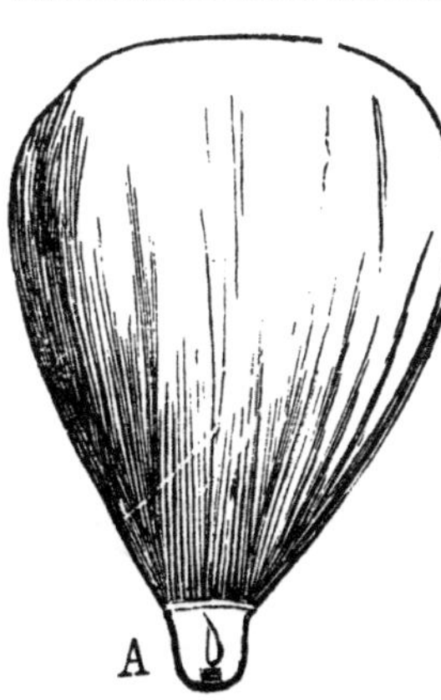

Fig. 30. AIR-BALLOON.

Because the *heat* of the flame, and the *burning of the spirit*, A, create a volume of *rarefied*, or *thin air*, which inflates the balloon, and makes it *specifically lighter* than the surrounding medium.

832. *Why do balloons sometimes burst when they ascend very high?*

Because, as they get into the *thinner air*, which exists at *high altitudes*, the gas within them expands, and the coating of the balloon is burst asunder.

833. *Why does the gas of balloons expand in thin air?*

Because the air exerts a *less amount of pressure* upon the air or gas contained in the balloons.

834. *Why do parachutes fall very gradually to the ground?*

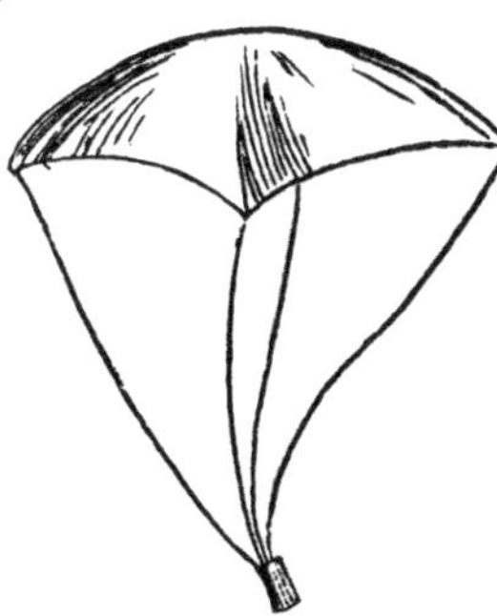
Fig. 31.—PAPER PARACHUTE.

Because the *air*, coming n contact with the *under surface* of the expanded head of the parachute resists its downward progress.

835. *Why does a shuttlecock travel slowly through the air?*

Because the air acts upon the feathers of the shuttlecock, in the same manner as it does upon the parachute—it strikes against their expanded surface, and resists their progress through the air.

836 *Why does the shuttlecock spin in the air?*

Because the surfaces of the feathers fall upon the air *obliquely*, or slantingly, and therefore, as the shuttlecock descends, it turns in the air.

FIG. 32.—BATTLEDORE AND SHUTTLECOCK.

837. *Why do we hear a noise when we strike the shuttlecock with the battledore?*

Because the *percussion* of the shuttlecock upon the parchment of the battledore causes it to vibrate, and the vibrations are imparted to the air.

838. *Why is the sound a dull and short one?*

Because the vibrations of the parchment are *not very rapid*, therefore there is *little intensity* in the vibrations of the air.

839. *Why does the exercise, afforded by playing battledore and shuttlecock, make us feel warm?*

Because it makes us breathe *more freely*, and causes the *blood to flow faster;* we, therefore, inhale more *oxygen*, which produces heat by combining with the *carbon* of our *blood.*

840. *Why does a kite rise in the air?*

A kite rises in the air by the force of the wind, which *strikes obliquely* upon its *under surface.* The string is attached to the "belly-band" in such a manner that it is nearer the *top* than the *bottom* of the band: this causes the bottom of the kite, when its surface is met by the wind, to recede in the direction of the wind the top is accordingly *thrown forward*, and the kite is made to *lie obliquely* upon the current of air moving against it. The kite then being *drawn by the string in one direction*, and *pressed by the air in another direction*, moves in a line which *describes a medium between the two forces acting upon it.*

841. *Why does the kite-string feel hot when running through the hand?*

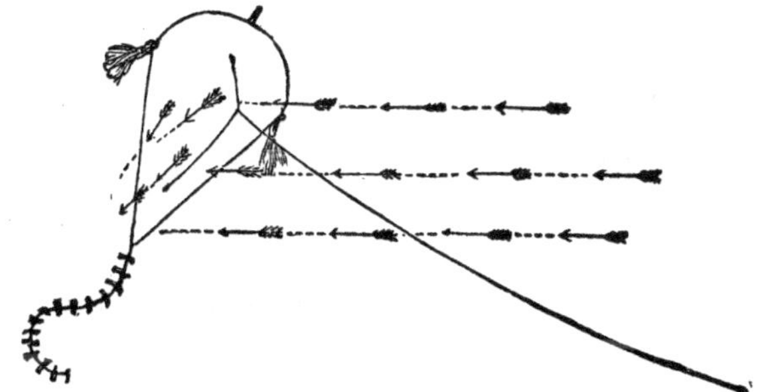

Fig. 33.—DIAGRAM EXPLAINING THE FLIGHT OF A KITE.

Because the *rapid friction* sets free the *latent heat* of the *string*, attracts the heat of the *hand* to the spot where the friction occurs, and sets free the latent heat of the *air*, which follows the *string* through the hand, and is compressed by the friction.

842. *Why does running with the kite cause it to rise higher?*

Because it *increases the force* with which the wind strikes upon the surface of the kite. If a person were to *run with a kite at the rate of five miles an hour, through a still air*, the effect would be *equal to a wind flying at the rate of five miles an hour* against a kite held by a *stationary string.*

843. *Why does the flying-top rise in the air?*

Because its wings *meet the air obliquely*, just as the surface of the kite does. And the *twirling of the top*, causing the oblique surfaces of its wings to strike the air, produces *the equivalent effect of a wind from the earth blowing the top upwards.*

844. *Why does the flying-top return to the earth when its rotations are expended?*

Because the *reaction* produced by its wings striking upon the air, is insufficient to counteract the *attraction of gravitation.*

Fig. 34.—FLYING-TOP.

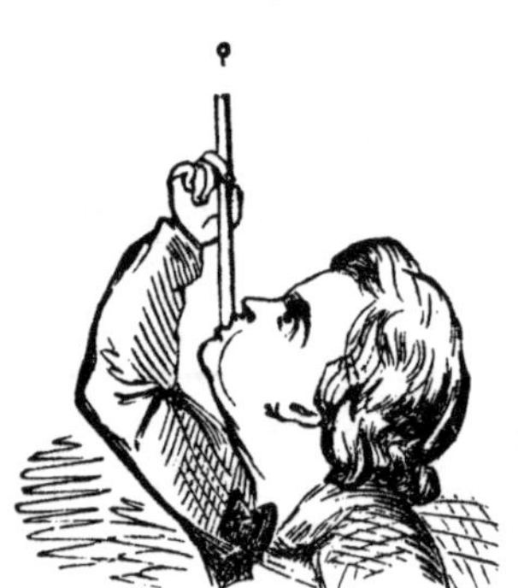

Fig. 35.—PEA AND PIPE.

845. *Why does a pea, into which a pin has been stuck, dance in suspension upon a jet of air blown through a pipe?*

Because the jet of air, being *slightly compressed* under the *convex* form of the pea, by the weight of the pin, forms a *concave cup of air,* in which the pea rests.

In the case put, it is supposed that the pin is *passed through the pea* until its head comes in contact with it. The pin is dropped into the hole of the pipe, and the breath is then applied, the pipe being held upright. The pea will rise in the air, and be suspended upon the jet, while the point of the pin will rotate around the stem of the pipe. There are other methods of fixing the pin which alter the result, and require a different explanation to that given above.

LESSON XL.

846. *Why does a mouse, painted upon one side of a card, and a trap upon the other, represent to the eye a*

"Honour thy father and thy mother * * That it may be well with thee, and thou mayest be long on the earth."—EPHESIANS VI.

mouse in a trap when the card is rapidly twirled upon a string?

Because the image of the mouse is brought to the retina of the eye before the image of the trap has passed away. The two impressions, therefore, ***unite upon the retina,*** and produce the image of a mouse in a trap.

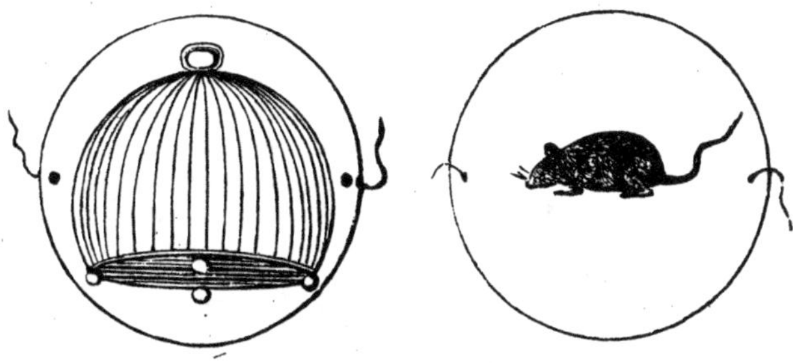

Fig. 36.—CARD WITH MOUSE-TRAP.

Fig. 37.—REVERSE OF CARD WITH MOUSE.

847. *Why will a bow stretched out of its natural position, propel an arrow through the air?*

Because its substance, being *highly elastic,* the particles thereof seek to restore themselves to their former state, as soon as the resisting power is withdrawn. The *force* derived from this elasticity, is communicated to the arrow by the string against which it is placed.

848. *Why is the arrow propelled forward?*

Because the elasticity of the bow, *acting equally upon its two ends,* to which the string is fastened, produce a line of force in a *diagonal direction.* It thus illustrates the law, that *when a body is acted upon by two forces at the same time, whose directions are inclined to each other, it will not follow either of them, but will describe a line between the two.*

849. *What forces tend to arrest the flight of the arrow?*

The *friction of the air*, and the *attraction of gravitation.*

"My son, give, I pray thee, glory to the Lord God of Israel, and make confession unto him."—JOSHUA VII.

850. *Why are feathers usually fastened to the ends of arrows?*

Because the *greater friction* of air acting upon them, opposes the progress of that part of the arrow in a greater degree than it does the other portion. The effect is, *to keep the point of the arrow forward*, and in a straight line with its opposite extremity. If the arrow were shot the reverse way from the bow, it would *turn round*, in the course of its flight, in consequence of the friction of the air, offering greater resistance to the progress of the feathered end.

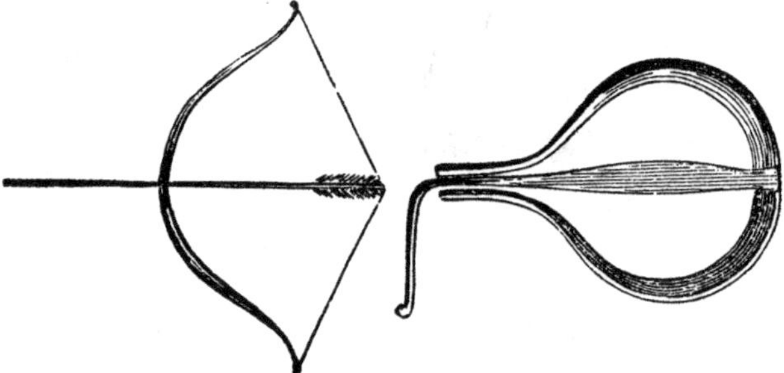

Fig. 38.—BOW AND ARROW. Fig. 39.—JEW'S HARP.

851. *Why does a Jew's harp give musical sounds?*

Because the *vibrations of the metal tongue* are communicated to the ear.

852. *Why will not the Jew's harp produce loud sounds unless it is applied to the mouth?*

Because the vibrations are not very intense, but when it is blown upon by the breath, the air is pressed upon it, and the vibrations are thereby rendered more powerful.

853. *Why does the alteration of the arrangement of the mouth, affect the formation of the sounds?*

Because it sends the air to the tongue of the harp in *a greater or lesser degree of compression.*

"Hear, ye children, the instruction of a father, and attend to know understanding."—PROVERBS IV.

854. *Why does the pressure applied to the handle of an air pistol propel the cork?*

Because, between the cork A and the air-tight piston C, there is a *closed chamber of air* B. When the handle D, which moves the piston C, is rapidly pushed in, it *compresses the air* until it is so much condensed, that it forces out the cork A.

Fig. 40.—AIR PISTOL, OR "POP-GUN."

855. *Why must the handle be drawn out, before the cork is placed in?*

Because otherwise a partial *vacuum* would be formed between A and C, and there would not be sufficient air to force out the cork by the return of the piston C D.

856. *Why does water rise in a syringe when the handle is drawn out?*

Because the pressure of the air on the water outside of the syringe, forces it into the space vacated by the drawing up of the handle, and where, otherwise, a *vacuum* would be formed.

Fig 41.—SYRINGE, WITH JET OF WATER.

857. *Why does not the water run out when the syringe is raised?*

Because the pressure of the air upon the small orifice resists the weight of the water.

858. *Why does the water leak out, but not run?*

Because water has a tendency always to *move to the lowest point,* but as the air does not enter freely the water cannot escape. It therefore *drops*, as small portions of the air enter.

859. *Why cannot the handle be pressed in, if the finger is applied to the orifice?*

Because water is not *compressible*, like air; it must therefore escape before the handle can be pressed in. Air may be forced into a much smaller compass than is natural to it; but it is impossible to *compress water* in any great degree.

Fig. 42.—"SUCKER." Fig. 43.—HOOP.

860. *Why does a "sucker" raise a stone?*

Because underneath the sucker *a vacuum* is formed and the external air, pressing on all sides *against the vacuum,* lifts the stone. The term "sucker" is founded upon the mistaken notion that the leather "sucks," or "draws" the stone. That such is not the case is evident: if, when the stone is suspended, a pin's point be passed under the leather, so as to open a small passage for the air, the stone will *drop instantly.*

861. *Why does a hoop roll, without falling to the ground?*

Because the *centrifugal force* gives it a motion which is called the *tangent to a circle*—that is, a tendency in all its parts *to fly off in a straight line.* When a piece of clay adhering to the hoop flies off, it leaves the hoop in a line which is straight with the part of the surface from which it was propelled; this line is *the tangent to the circle of the hoop;* and the tendency of all the parts of the hoop to fly off in this manner, counteracts the attraction of the earth, so long as the hoop is kept in motion.

862. *Why does the hoop, in falling, make several side revolutions?*

Because its onward movement, not being quite expended, influences the *centre of gravity of the hoop*, and changes its line of direction. The hoop is also elastic, and when its sides strike the earth, they spring up again, and continue turning until the opposing forces are overcome by the *attraction of gravitation.*

863. *Why will a little boy balance a large boy on a see-saw?*

Because the "see-saw" may be placed so that its ends are at *unequal distances from the centre.* This gives the little boy the power of *leverage*, by which is meant the increase of power, or weight, by *mechanical means.*

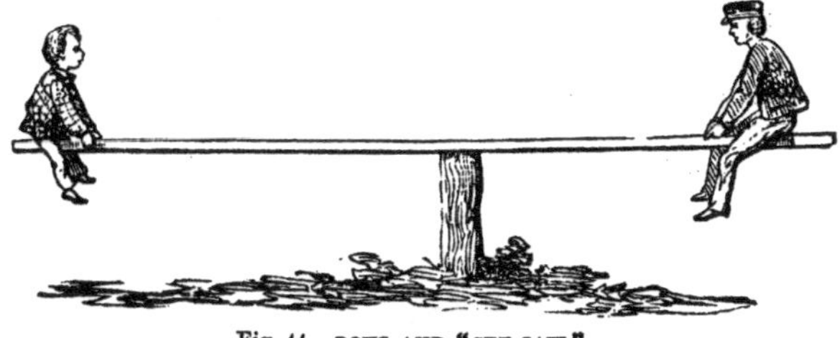

Fig. 44.—BOYS AND "SEE-SAW."

864. *Why does the little boy sink to the ground when the larger boy slightly kicks the earth?*

Because the larger boy, by kicking against the earth, opposes by mechanical force the *attraction of gravitation* acting upon him, and he becomes *temporarily* less attracted to the earth than the little boy.

865. *Why can the little boy, if he choose, keep the big boy up, when once he is up?*

"Little children, let no man deceive you: he that doeth righteousness is righteous, even as he is righteous."—1 JOHN III.

Because, as the big boy is then on *an inclined plane* with the *fulcrum*, or centre upon which the see-saw moves, the arm of *the lever*, upon which the big boy sits, is *relatively shortened*, and he has then *less mechanical power*. Also, a portion of the weight of the larger boy is transmitted along the lever *to the arm upon which the little boy sits.*

866. *Why is the ball propelled upward, in the game of trap and ball, when the trigger is struck?*

Because, when the trigger is struck at A, it is forced downwards, turning upon the fulcrum B, the opposite end, forming the spoon, is thereby forced upwards, describing a small *arc*, or curved line; but directly the ball is set free from the spoon, it rises in a *right line* with the *direction it was taking, at the moment it was set free.*

Fig. 45.—TRAP AND BALL.

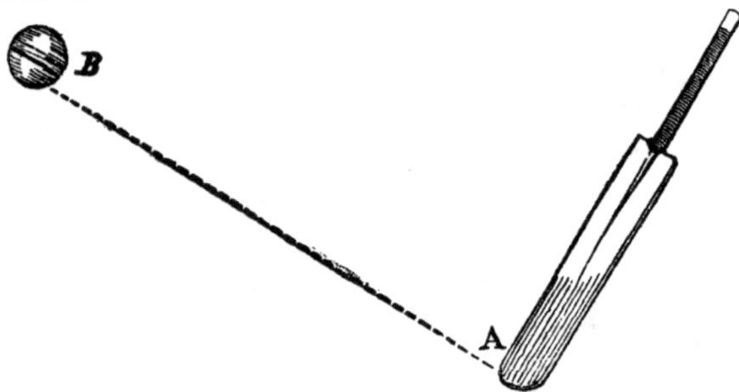

Fig. 46.—BAT AND BALL.

867. *What principles of natural philosophy are illustrated by the results of bat and ball?*

Percussion, when the bat strikes the ball; *rotatory motion*, when the ball is sent whirling away; *momentum*, which it acquires by velocity; *elasticity*. when it rebounds from an object against which it strikes; *reflected motion*, when it is turned by a body upon which it impinges; *friction*, as it rolls along the ground; the *communication of force*, when it sets another body in motion against which it strikes; *gravitation*, when it falls to the earth; and *inertia*, when it lies in a state of rest.

868. *Why do pith-tumblers always pitch upon one end?*

Fig. 47.
PITH-TUMBLER.

Because the *lead* B is *specifically heavier* than the *pith* to which it is attached; it therefore always falls undermost; and as the lead is rounded off, just like the spill of a top, after the head has oscillated a little, and expended the force of the momentum of its fall, it will settle upon its *centre of gravity*, or the point through which it is *attracted to the earth.*

869. *Why do the figures upon the "Thaumatrope" appear to dance, when they are made to revolve before a mirror?*

Because the eye, in looking through the holes in the card, towards the reflections in the mirror, receives *a rapid succession of impressions.* As the figures upon the card are represented in a graduated series of positions—the *first* one standing upright, the *second* with his knees a little bent, the *third* a little more bent, as in the act of springing, and so on, the *figure* being in each case *the same*, but the position *slightly altered*, imparts an impression to the mind, through the eye, that *one figure* is passing through a *series of motions.*

Thaumatrope.—From two Greek words, meaning *wonder* and *to turn.*

We have said enough, we hope, to show that even the play-hours of children may be made instructive to them; and that the simplest toys may be used to illustrate some of the grandest laws of nature. Nor may this kind of instruction be confined to children alone. Grown-up people, whether participators in the sports of youth, or simple observers of their games, may gain instruction for themselves, and be the better teachers of their children, by taking an interest in their enjoyments, and giving to their minds, through the attractiveness of pastime, a taste for observing and estimating the varied phenomena which present themselves.

Moreover, we think that parental government acquires a greater power when

It leans towards the natural desires of childhood, and wins those desires into a proper direction. Love existing between parent and child is the best tie to home, and the strongest incentive to duty. There is also something in the gentleness of childish nature which may influence for good the sterner mould of man too often warped and clouded by the cares of life.

Fig. 48.—THAUMATROPE, OR "WONDER-TURNER."

In Kay's "Life of Sir John Malcolm," we find an admirable and apt passage. Sir John says:—"I have been employed these last few hours with John Elliot and other boys, in trying how long we could keep up two cricket-balls. Lord Minto caught us. He says he must send me on a commission to some very young monarch, for that I shall never have the gravity of an ambassador for a prince turned of twelve. He, however, added the well-known and admirable story of Henry IV. of France, who, when caught on all fours carrying one of his children, by the Spanish envoy, looked up and said, 'Is your excellency married?' 'I am, and have a family,' was the reply. 'Well, then,' said the monarch, 'I am satisfied, and shall take another turn round the room,' and off he galloped, with his son on his back flogging and spurring him. I have sometimes thought of breaking myself of what are termed boyish habits; but reflection has satisfied me that it would be very foolish, and that I should esteem it a blessing that I can find amusement in everything, from tossing a cricket-ball, to negotiating a treaty with the Emperor of China. Men who will give themselves entirely to business, and despise (which is the term) trifles, are very able, in their general conception of the great outlines of a plan, but they feel a want of knowledge, which is only to be gained by mixing with all classes in the world, when they come to those lesser points upon which its successful execution may depend."

"Whether therefore ye eat, or drink, or whatsoever ye do, do all to the glory of God."—CORINTH. X.

CHAPTER XLI.

*869. *Why do we eat food?*

Because the atoms of which our bodies are composed are ***continually changing.*** Those atoms that have fulfilled the purposes of nature are removed from the system, and, therefore, new matter must be introduced to supply their place.

870. *Why do we eat animal and vegetable food?*

Because their substances are composed of ***oxygen, hydrogen, carbon,*** and ***nitrogen***—the four chemical elements of which the human system is formed. They are, therefore, capable of nourishing the body, after undergoing digestion.

871. *Why do we masticate our food?*

Because mastication is *the first process towards the digestion of food.* Before animal or vegetable substances can nourish us, their condition must be entirely changed, their *organic* states must be dissolved, and they must become simple matter, in a homogeneous mass, consisting of the four chemical elements necessary to nutrition, and they must again be restored to an organic condition.

872. *Why does saliva enter the mouth when we are eating?*

Because, in addition to the *mechanical* grinding of the food by the action of the teeth, it is necessary that it should undergo certain chemical modifications to adapt it to our use. There are placed, therefore, in various parts of the body, *glands,* which secrete peculiar fluids, that have a chemical influence upon the food.

The first of these glands are the *salivary glands of the mouth.* which pour out a clear watery fluid upon the food we eat, and which fluid has been found to possess a property which contributes to the digestion of food.

The moisture afforded by the salivary secretion is also necessary to enable us to swallow the food.

873. *Why does the salivary juice enter the mouth just at the moment that we are eating?*

"And the Lord said unto him, Who hath made man's mouth? or who maketh the dumb, or the seeing, or the blind? have not I the Lord?"—EXODUS IV.

Because the glands, which are buried in the muscles of the mouth, and which in their form are much like bunches of currants are always full of salivary secretion. There are nerves which are distributed from the brain to these glands, and when other nerves which belong to the senses of taste, of sight, or of feeling, are excited by the presence of food, *a stimulus* is imparted to the salivary glands, through the nerves that surround them, their cells collapse, and the juice which they contain is poured out through their stems, or ducts, into the mouth.

874. *How do we know that impressions imparted to one set of nerves, may be imparted to another set, so as to put any particular organ in action.*

Because very frequently *the mere sight* of rich fruit, or acid substances, *will cause the saliva to flow freely.* In this case it is evident that the salivary glands *could not see or know* that such substances were present. An impression must, therefore, be made upon the brain, *through the organ of vision,* and the desire to taste the substances being awakened, a nervous stimulus is *imparted to the glands of the mouth,* and they at once commence their action, *as if food were present.*

875. *Why does food descend into the stomach?*

Because, after the teeth, the tongue, and the muscles of the mouth generally, have rolled the food into a soft bolus, it is conveyed to the back of the mouth, where it is set upon the opening of the throat (*œsophagus*). It does not then descend through the throat by its own gravity, because the throat is generally in a compressed or collapsed state, like an empty tube; and we know that persons can eat or drink when with their heads downwards. The œsophagus is formed of a number of muscular threads, or rings, and *each little thread is like a hand ready to grasp at the morsel that is coming.* As soon as the bolus is presented at the top of the throat, these little muscular hands lay hold of it, and transmit it downwa d, passing it from one to another, until it is conveyed through the long passage, to the door of the stomach, which it enters.

"Remove far from me poverty and lies; give me neither poverty nor riches; feed me with food convenient for me."—PROVERBS XXX.

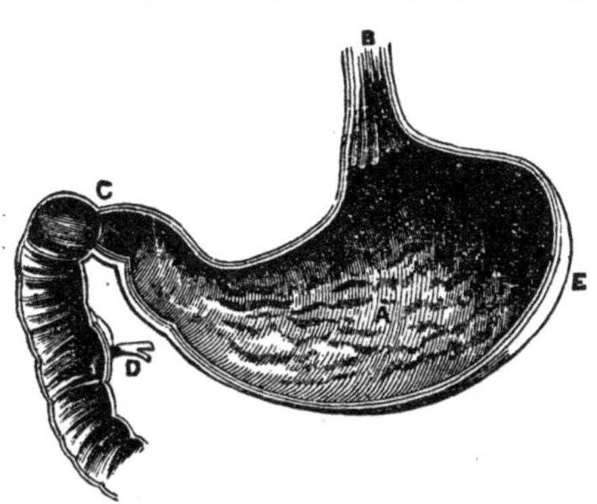

Fig. 49.—SECTION OF THE STOMACH, &c.

A. The inner coat of the *stomach*. (The stomach is here represented cut through its length, so that we can see its inside.)

B. The lower extremity of the throat, or *œsophagus*, through which food enters the stomach.

C. The passage out of the stomach, called the *pylorus*, where a muscular contraction prevents the escape of undigested food.

D. The *duodenum*, and the ducts through which the *bile* and *pancreatic* juices enter and mingle with our food.

876. *Why do we not feel the food being transmitted through the throat?*

Because the nerves of the body differ in their powers: some are nerves of *feeling*, some of *motion*, and others are nerves of the *senses*. The nerves of feeling are most abundantly distributed to those parts *where feeling is most useful and necessary to us*. But the faculty of feeling our food undergoing digestion would be no service to us whatever; therefore the nerves of *motion* are plentifully distributed to the throat and stomach, but very few of the nerves of *feeling*—just as many as will tell us when we eat anything *too hot*, or *too cold*, or that the stomach is *out of order*.

877. *Why do we feel uneasy after eating to excess?*

Because the stomach is *distended*, and presses upon the other organs by which it is surrounded.

"Who satisfieth thy mouth with good things; so that thy youth is renewed like the eagles."—PSALM CIII.

878. *Why do we feel drowsy after eating heartily?*

Because, while the stomach is in action, *a great proportion of the blood of the body is drawn towards it,* and as the blood is withdrawn from the other parts of the body, they fall into a state of languor.

879. *Why does blood flow more freely to the stomach during digestion?*

Because the energy of an organ is *increased by the flow of blood,* which supplies the *material* of which our organs are composed, and in which the *vital essence,* supporting life, resides.

880. *Why does excess in eating bring on indigestion?*

Because the power of the stomach to digest food is *governed by the amount of food required by the system.* It seems to be an instinct of the stomach to hold back food which is in excess, and by indications of pain and disturbance to warn its master that *excess has been committed.*

881. *Why is food digested in the stomach?*

Because it enters the stomach in the form of a paste, produced by the action of the mouth; and directly food enters, the *gastric juice,* which is formed by glands embedded in the coats of the stomach, trickles down its sides. This is a more *powerful solvent* than the salivary juice—it is like the same kind of fluid, only much stronger, and it soon turns the food from a rough and crude *paste* into a *greyish cream* (chyme). The heat of the stomach assists the operation, and the muscular threads of the coats move the cream along, in the same manner that the muscles of the œsophagus brought down the food.

The cream is passed towards the door which leads outward from the stomach (*pylorus*); but if, in the midst of the cream, there are any undissolved particles of food, it closes upon them, and they return again to the stomach to be further changed.

882 *Why does indigestion bring on bilious attacks?*

Because the *liver* secretes a fluid to assist in the digestion of food. The liver is a gland—a similar organ to the glands of the mouth—and it forms *bile* in the same manner that they form the salivary juice. Only the liver is a *much larger gland*, and a much greater quantity of blood passes through it. The liver pours its secretion into the biliary duct (Fig. 49) to mix with the grey cream as it passes onward, and to further dissolve it. But when the stomach is excited by food which it cannot dissolve, and when the owner of the stomach, disregarding its remonstrances, will persist in over-eating, or in eating things that disagree with the system, then *the liver and the stomach sympathise*, and the muscular threads, or hands, that prevail all through the alimentary organs, instead of moving *onward*, move backward, and *throw some bile into the stomach* to assist to dissolve and remove the excessive or improper food.

CHAPTER XLII.

883. *Why does some portion of the food we eat nourish the system, while other portions are useless?*

Because most food contains some particles that are indigestible, or that, if digested, are innutritious, and not necessary for the system. The *liver* is the organ by whose secretion the *useful is separated from the useless;* for when the bile enters through the duct (Fig. 49) and mixes with the grey cream coming from the stomach, it remains no longer a grey cream, but turns into a mass coloured by bile, having upon its surface *little globules of milk*, small, but very white. Those minute globules of milk (*chyle*) are the nutritious particles derived from the food; the other portion, coloured with bile, is the useless residue, or rather the *bulk from which the nutrition has been extracted.*

884. *Why does the milky, or nutritious matter, separate from the innutritious, upon admixture with bile?*

Because the bile contains an oily matter which *repels* the *watery milk of nutrition.*

"God hath made of one blood all nations of men for to dwell on all the face of the earth."—ACTS XVII.

The *pancreatic juice* also enters through the same duct with the bile. But its precise use is not understood. It is a fluid much like the salivary secretion of the glands of the mouth.

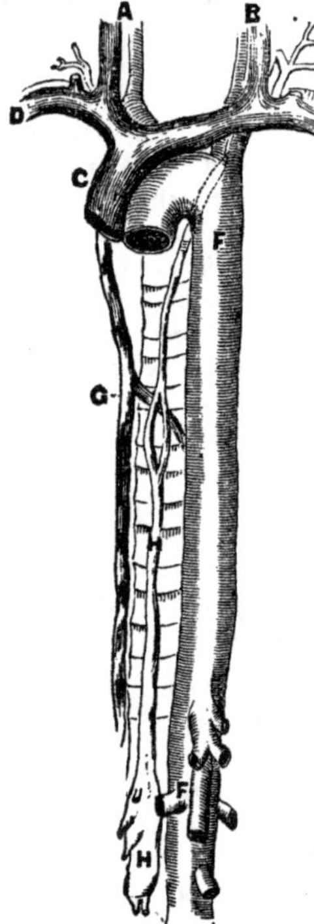

Fig. 50.—GREAT VESSELS OF THE CIRCULATION, AND THE DUCT WHICH CONVEYS NUTRITIVE MATTER TO THE BLOOD.

A B. *Jugular veins* which return blood from the head to the heart.

C. The *superior venæ cava*, or trunk vein, which pours the blood returned from the upper part of the system into the heart. There is a similar large vessel which meets this one and brings back blood from the lower part of the body, and they both pour the blood into the right side of the heart.

D E. The branches of the *venous system* which bring back the blood from the arms.

F F. The *great aorta*, the blood vessel which conveys arterial blood from the heart, and gives off branches that supply every part of the body.

G. Another large vein which returns the blood from the muscles of the chest, &c.

H H. The *thoracic duct*, which receives the newly dissolved food from the small absorbents, that collect it from the intestines. It conveys this nutrition (called chyle) upward along the back, until it reaches where the duct turns into the junction of two veins, and pours its contents into the veins bringing blood back to the heart. The nutrition, therefore, is at this moment mixed with the venous blood, and is sent to the lungs to be oxygenised.

885. *How is the nutrition taken away from the bilious residue?*

The muscular threads (or hands, as we figuratively call them) continue to push forward the digested matter through a long tube, called *the alimentary canal*, or bowels. This canal is some thirty feet in length, and is folded in various layers across the abdomen, and tied to the edge of a sort of apron, which is gathered up and fastened to the back-bone. All along this alimentary canal those muscular bands are pushing the digested mass along. But upon the coat or surface of the canal there are millions of little vessels called *lacteals*, which look out for the minute globules of milk as they pass, and *absorb* them, which means that they pick them up, and carry them away. There is an immense number of these little vessels, all busily at work picking up food for the system.

Then there is a large vessel, called the *thoracic duct*, which comes down and communicates with those little vessels (it is a sort of overseer, having a large number of workmen,) and collects the produce of their toil, and carries it upwards to the part where it passes *from the organs of digestion* into the *vessels of circulation.*

886. *What becomes of the nutrition, when it has entered the vessels of the circulation?*

It is sent through a large vein into *the heart*, entering that organ on the right side, from which the heart propels it into the lungs, mixed with *venous blood;* and the venous, or blue blood, is sent into the lungs, *taking with it the milk*, the formation of which we have traced.

887. *Why are the venous blood and the chyle sent to the lungs?*

Because the venous blood, in its circulation through the body, has parted with its *oxygen*, and taken up *carbon*, and it requires *to get rid of the carbon, and take up more oxygen.* The chyle, also, now combined with the blood, requires *oxygen*, and having obtained it, is converted into *bright red blood*, and the blue blood of the veins, having got rid of its carbon, which formed the carbonic acid of the breath, has again become *bright red blood.* We must therefore, in pursuing our description, cease to speak of blue, or *venous blood*, and of white milk, or *chyle*, for the two have now combined, and, with the oxygen of the air, have formed *arterial blood.*

"My flesh and my heart fainteth; but God is the strength of my heart, and my portion for ever."—PSALM LXXIII.

888. *What becomes of the arterial blood thus formed?*

It is sent back from the lungs to the right side of the heart, from which it is sent into the *great trunk of the aorta*, and from thence it passes into smaller blood-vessels, until it finds its way to *every part of the system.*

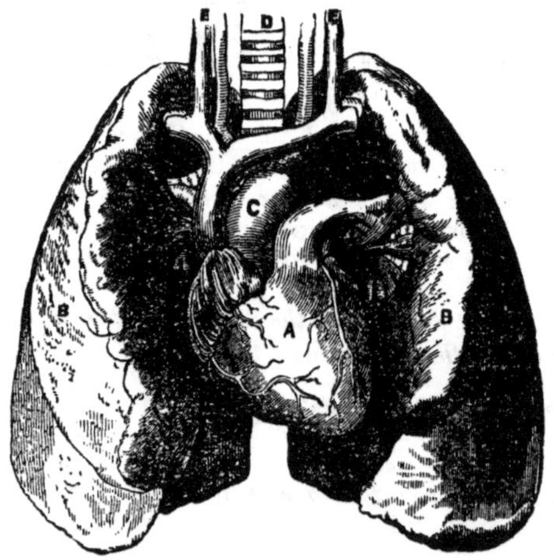

Fig. 51.—THE ORGANS OF RESPIRATION.

A. The *heart.*

B B. The *lungs.*

C. The *aorta*, and on either side of the aorta the vessels which convey the venous blood to the lungs to be *oxygenised*, and the corresponding vessels which return it to the heart, after it has undergone that operation. (For *aorta* see Fig. 50.)

D. The *trachea*, or large air passage, through which the air passes into the spongy texture of the lungs, when we breathe.

E E. *Arteries* and *veins*, being the trunks of the vessels that supply the head, &c.

889. *Why does the chest expand when we breathe?*

Because the lungs consist of *millions of hollow tubes*, and *cells*, which, having been emptied by throwing off *carbonic acid gas* and *nitrogen*, become compressed, and the atmospheric air flowing into these millions of spaces, and filling the lungs, just as water fills and swells a sponge, causes them to expand, and occupy greater room.

890. *How does the blood communicate with the air in the lungs?*

Through the *sides of very minute vessels*, of which, perhaps, a *fine hair* gives us the best conception. But these vessels are *twisted and wound round each other* in such a curious manner, that they form *millions of cells*, and by being twisted and wound, a much *greater surface of air and blood* are brought to act upon each other, than could otherwise be accomplished.

891. *Why does the blood which is thus formed, impart vitality to the parts to which it is sent?*

Because the blood is itself *vitalised*—is, in fact, *alive*, and capable of diffusing life and vitality to the organisation of which it forms a part.

This is a very wonderful fact, but no less true than wonderful, that dead matter which, but a little while ago, was being ground by the teeth, softened by the saliva, and solved by the gastric juice and bile, has now acquired *life*. Nobody can tell the precise stage or moment when it began to live. But somewhere between the stomach and the lungs, melted by the gastric juice, softened by the secretion of the pancreas, separated by the bile of the liver, macerated by the muscular fibres of the bowels, taken up by the absorbents, warmed by the heat of the body, and ærated in the lungs, it has by one, or by all of these processes combined, been changed from the dead to the living state, and now forms part of the *vital fluid of the system*.

CHAPTER XLIII.

892. *Why do we know that the blood has become endowed with vital powers?*

Because, in the course of its formation, it has not only undergone change of condition and colour; but, if examined now by the microscope, it will be found to consist of millions of minute cells, or discs, which float in a watery fluid. The paste produced by mastication consisted of a crude admixture of the atoms of food; the cream (*chyme*) formed from this in the stomach, presents to the microscope a heterogeneous mass of matter, exhibiting no appearance whatever of a new organic arrangement; the milk (*chyle*) which is formed in the intestines is found to contain a great number of very small molecules, which probably consist of some fatty matter; as the chyle progresses towards the *thoracic duct* (Fig. 50), it appears to contain more of these, and slight indications present themselves of the approach towards a new organic condition.

But wherever *vitalisation begins*, no human power can say with confidence. Yet there can be no doubt that the blood is both *organised* and *vitalised*, and that it consists of corpuscles, or little cells, enclosing matters essential to life.

893. *Why does the blood circulate?*

Because all the bones, muscles, blood-vessels, nerves, glands, cartilages, &c., of which the body is composed, are constantly undergoing a *change of substance*. It is a condition of their life, health, and strength, that they shall be "*renewed*," and the blood is the great source of the *materials* by which the living temple is kept in repair.

894. *How is the body renewed by the blood?*

Every drop of blood is made up of a large number of corpuscles, each of which contains some of the elements essential to the wants of the system.

Let us, to simplify the subject, consider the blood vessels of the body to be so many *canals*, on the banks of which a number of inhabitants live, and require constant sustenance. The corpuscles of the blood are the *boats* which are laden with that sustenance, and when the heart beats, it is a signal for them to start on their journey. Away they go through the arch of the great *aorta*, and some of the earliest branches which it sends off convey blood to the arms. We will now for a moment dismiss the word *artery*, and keep up the figure of a system of canals, with a number of towns upon their banks.

Well, away go a fleet of boats through the *aorta* canal, until they reach a point which approaches Shoulder-town; some of the

"Though hand join in hand, the wicked shall not be unpunished; but the seed of the righteous shall be de ivered."—PROVERBS XXI.

boats pass into the *axillary* canal and Shoulder-town is supplied; the other boats proceed along the *humeral* canal until they approach Elbow-town, when another division of the boats pass into other branch canals and supply the wants of the neighbourhood; the others have passed into the *ulnar* canals and the *radial* canals until they have approached Wrist-town and Hand-town, which are respectively supplied; and then the two canals have formed a junction across the palm and supplied Palm-town, where they have given off branches and boats to supply the four Finger-towns, and Thumb-town.

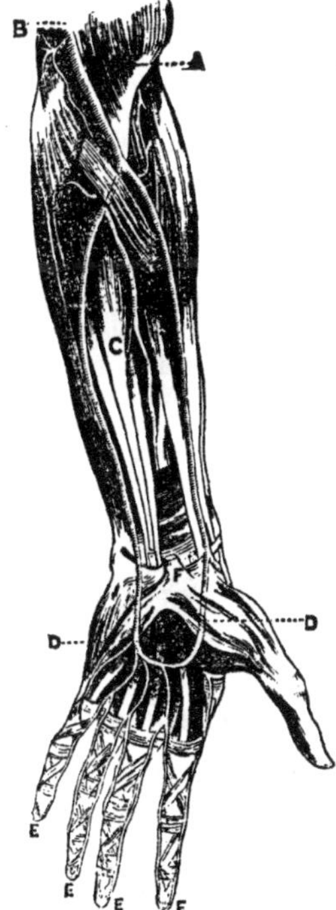

FIG. 52. — ILLUSTRATION OF THE SYSTEM OF CANALS THAT SUPPLY THE FORE-ARM WITH BLOOD.

Between A and B the *brachial canal*, which gives off branches to supply Elbow-town, &c., and then divides into two main courses, diverging to the opposite sides of the arm, and sending a smaller canal down the centre.

D D. The point where the *ulnar canal* and the *radial canal*, after having passed and supplied Wrist-town, form a junction, running through Palm-town, and in their course giving off branches to supply the four Finger-towns and Thumb-town.

For further explanations of the engraving, see 57.

895. *How does the blood return to the lungs, after it has reached the extremities?*

The *veins* constitute a system of vessels corresponding to the arteries. We may say that the arteries form *the down canal*, and the veins *the up canal.* The arteries, commencing in the great trunk of the *aorta*, branch off into large and then into smaller tubes, until they form *capillary* or hair-like vessels, penetrating everywhere.

The capillary extremities of the arteries, unite with the capillary extremities of the veins, and the blood passes from the one set of vessels into the other. As the *arteries become smaller* from the point where they receive the blood, so the *veins grow larger*; the venous capillaries, pour their contents into small vessels, and these again into larger ones, until the great *venous* trunks are reached, and the blood is passed again into the heart as at first described. (Fig. 50.)

896. *Why do we see blue marks upon our arms and hands?*

Because large veins lie underneath the skin, through which the blood of the fingers and hand is *conveyed back to the heart.*

897. *Why are the veins more perceptible than the arteries?*

Because the arteries are buried *deeper in the flesh, for protection.* It would be *more dangerous to life to sever by accident an artery than a vein.* A person might bleed longer from a vein than from an artery, without endangering life; because the arteries supply the *life sustaining blood.* The Almighty, therefore, has buried the arteries for safety.

898. *Why when we prick the flesh with a needle does it bleed?*

Because the capillary arteries and veins are so fine, and are so thickly distributed all over the body, that not even the point of a needle can enter the flesh without penetrating the coats of several of these small vessels.

899. *What occurrs during the circulation of the blood?*

Not only do the various parts to which the boats are sent take from them whatever they require, but *the boats collect all those matters for which those parts have no further use.* The bones, the nerves, the muscles, &c., all renew themselves as the boats pass along; and all give something to the boats to bring back. One of the chief exchanges is that of *oxygen* for *carbon*, by which a gentle *heat* is diffused throughout the system. It is for this purpose that *fresh air* is so *constantly necessary*.

But other exchanges take place. The blood, in addition to oxygen and carbon, contains *hydrogen* and *nitrogen*. But it contains its four elements in *various forms of combination*, producing the following *materials* for the use of the body : of 1,000 parts of blood, *about* 779 are *water*; 141 are *red globules*; 69 are *albumen*; 3 are *fibrin*; 2 are *fatty matter*; 6 are various *salts*.

Albumen and fibrin are a kind of flesh imperfectly formed, and probably are chiefly used in repairing the muscles. The red corpuscles contain the oxygen which goes to combine with the superabundant carbon, and develope heat; the fatty matters probably repair the fatty tissues, and glands that are of a fatty nature; and the various salts contribute to the bones, and to the chemical properties of those secretions which are formed by the glands, &c., while the great proportion of water is employed in cleansing, softening, and cooling the whole, or the living edifice, and it is the medium through which all the nutrition of the body is distributed.

900. *Why do we feel the pulse beat?*

Because every time that the heart contracts it send a fresh supply of blood to the blood-vessels, and the motion thus imparted creates *a general pulsation throughout the system*: but it is more distinctly perceived at the pulse, because there a rather *large artery lies near to the surface*.

901. *What becomes of the matter collected by the blood in the course of its circulation?*

We have already explained that carbon is thrown off from the lungs in the form of carbonic acid gas. But there are many other matters to be separated from the venous blood, and its putrification is assisted by the action of the liver, which is supplied with a large vein, called the *portal vein*, which conveys into the substance of the liver, a large proportion of the venous blood, from which that organ draws off those matters which form the bile, and other matters which are transmitted with the bile to the bowels. The *liver* and

"Thy hands have made me and fashioned me: give me understanding, that I may learn thy commandments."—PSALM CXIX.

the *lungs*, therefore, are the great purifiers of the venous blood But there are also smaller organs that assist in the same work.

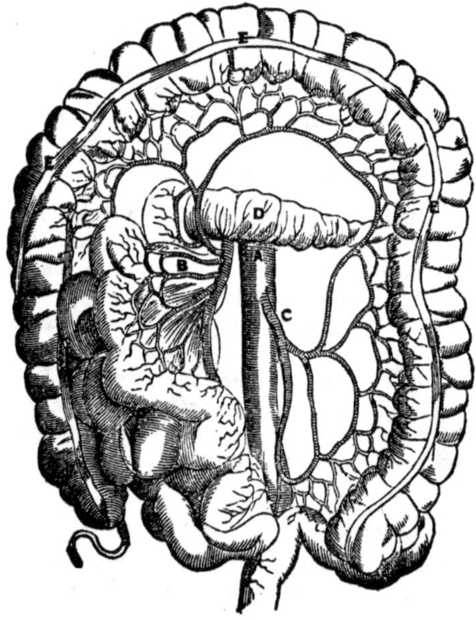

Fig. 53.—SHOWING THE DISTRIBUTION OF BLOOD THROUGH BRANCHES OF THE AORTA.

A. The *aorta*.

B. Branches given off for the *aorta* to supply one portion of the intestines.

C. Branches given off by the aorta to supply other portions of the intestines. A complete communication may be traced between these vessels from the origin of one to that of the other.

D. The *pancreas*, or sweetbread, a large gland that forms the pancreatic juice, which it pours in through the duct. *See* Fig. 50.

E E E. The *large intestines*, forming the termination of the alimentary canal.

CHAPTER XLIV.

902. *Why when we cut our flesh does it heal?*

Because the blood coagulates over the cut, and throws out a kind of *lymph*, which forms an incipient flesh, and excludes the air while the blood-vessels are engaged in *repairing the part.*

903. *Why, since all the substance of the body undergoes change, do we preserve the same features throughout our lives?*

Because our substance changes in the *minutest atoms;* and each separate atom has a life of itself, the maintenance of which preserves the *unity and permanence of the whole.*

904. *Why do moles upon the skin continue permanent, while bruises and wounds disappear?*

Because moles are themselves *organised formations*, and repair themselves just as any other part of the body does. But bruises and wounds are the result of *accidental disturbances*, which in course of time become removed.

905. *Why do the marks of deep cuts sometimes remain?*

If the cut is so deep and serious as to destroy the *system of vessels* which supply and repair the part, then it is evident that they cannot work so perfectly as when in their sound condition. Their functions are, therefore, interfered with, and instead of having flesh uniform with the other parts of the system, there results a *scar*, or a wound *imperfectly repaired.*

906. *Why when we hold our hands against a candle light do we perceive a beautiful crimson colour?*

Because the fluids and vessels of the body are in some degree transparent, and the thin textures of the sides of the fingers allows the light to pass, and shows the beautiful crimson colour of the blood.

If the web of a frog's foot be brought in the field of a good microscope, and set against a strong light, the blood may be seen in circulation, with the most wonderful effect. Each vessel, and every globule of blood, can be seen most distinctly, and the junction of the arteries and veins can be clearly traced. The little boats of nutrition may be seen chasing each other in rapid succession, and when the animal exerts itself to escape, the flow of the blood increases; and not unfrequently, under these circumstances of agitation, have we seen two or three blood discs struggling together to enter a vessel that was too small for them. Again and again they have endeavoured to find a passage, until one of them happening to slip forward, got away, followed by the others!

"Know ye that the Lord he is God: it is he that hath made us, and not we ourselves: we are his people, and the sheep of his pasture."—PSALM C.

907. *Why does the flesh underneath the nails look red?*

Because the transparent texture of the nails enables us to see the colour of the *vascular structure* that lies underneath the skin.

Vascular.—Full of vessels. In this instance, full of capillary blood-vessels.

908. *Why have we nails at our fingers' ends?*

Because they give *firmness to the touch,* and enable us to apply the extremities of the fingers to many useful purposes for which they would otherwise be unfitted. They enable us to *press the tips of the fingers,* where the highest degree of sensitiveness prevails, so as to bring *the largest amount of nervous perception* into the sense of touch.

909. *Why do white spots occur upon the nails?*

Because the vascular surface underneath is attached to the horny texture of the nail; but by knocks and other causes, the nail sometimes *separates in small patches from the membrane* below, and becomes *dry and opaque.*

910. *Why is there a circular line of whitish colour at the root of the nail?*

Because there the nail is *newly formed* by the vascular substance out of which it grows, and has not yet assumed its proper horny and transparent nature.

911. *Why is the eye-ball white?*

Because the blood-vessels that supply its surface are so very fine that they do not admit the *red corpuscles* of the blood.

912. *Why does the eye-ball sometimes become blood-shot?*

Because, under exciting causes of inflammation, the *blood-vessels become distended,* and the red corpuscles enter, producing a network of red blood-vessels across the white surface of the eye.

913. *Why are the lips red?*

Because the lips are formed of the *mucous membrane* that lines the body internally, and covers the surface of most of the internal parts. This membrane contains a great number of minute red vessels, which give softness and moisture to the surface. A very beautiful illustration of the softness, moisture, and delicate colour of the mucous membrane is afforded by turning up and examining the under surface of the upper eye-lid.

914. *Why do delicate persons look pale and languid?*

Because, generally from the want of exercise and fresh air, their blood is deficient of the healthy proportion of *red corpuscles.*

915. *Why does exercise and fresh air impart to healthy persons a red and fresh appearance?*

Because the redness of the blood is due to the *amount of oxygen* which it contains, and air and exercise *oxygenise* the blood, and diffuse it throughout the system.

916. *How is the blood propelled through the arteries?*

By the very powerful contraction (and alternate dilation) of the thick *muscles of the heart,* assisted also by the *muscular cords of the blood-vessels* themselves, and in many instances by the *compression of the muscles* in which the arteries lie embedded.

917. *Why are the capillary arteries capable of receiving the great quantity of blood sent out through the larger vessels?*

Because the capillary vessels are *so numerous,* that though they are infinitely smaller, they are capable of receiving in their minute tubes *the whole of the quantity of blood* transmitted to them through the larger vessels.

918. *Why, when we sit with our legs crossed, do we see the foot that is raised move at regular intervals?*

Because the pressure upon the muscles of the leg retards the progress of the blood until it forces *itself through the compressed vessels,* and thereby imparts a pulsation which moves the leg and foot.

919. *Why are capillary blood-vessels found in every part of the system?*

Because it is *through these small vessels alone* that the substances of the body are renewed and changed. Even the larger blood-

"All my bones shall say, Lord, who is like unto thee, which deliverest the poor from him that is too strong for him, yea, the poor and the needy from him that spoileth him?"—PSALM XXXV.

vessels *do not sustain themselves upon the blood which they contain* but receive into their coats numerous capillary vessels by which they are nourished.

920. *How much blood does the human body contain?*

From *twenty-five* to *thirty-five* pounds. (*See* 623.)

921. *How does the blood ascend in the veins, in opposition to gravitation?*

In addition to the muscular coats of the veins, and the influence of muscular action upon them, there are in the veins numerous semi-circular valves, which are not found in the arteries. These valves extend from the sides of the veins in such a manner that they allow the free passage of the blood upwards, but a backward motion of the blood would expand the cup-like valves and stop the passage: so that the blood can only move in one direction, and that *towards the heart.*

922. *How frequently does the total amount of blood circulate through the system?*

The blood circulates once through the body in about *two minutes.* If, therefore, we estimate the amount of blood at twenty-four pounds, it follows that no less than *twelve pounds of blood pass through the heart every minute;* and it is estimated that if the blood moved with equal force in a straight line it would pass through *one hundred and fifty feet in a minute.*

CHAPTER XLV.

923. *How many bones are there in the human body?*

There are *two hundred and forty-six*, and they are apportioned to the various parts of the body in the following numbers:—

Head	8
Ears	6
Face	14
Teeth	32
Back-bone and its base	26
Chest, &c.	26

"Our bones are scattered at the grave's mouth, as when one cutteth and cleaveth wood upon the earth."—PSALM CXLI.

Arms and Hands	64
Legs and Feet	62
Small moveable bones	8

924. *Of what substances are the bones composed?*

One hundred parts of bone consist of

Cartilage	32·17	parts
Blood-vessels	1·13	„
Carbonate of lime	11·30	„
Phosphate of lime	51·04	„
Fluate of lime	2·00	„
Phosphate of Magnesia	1·16	„
Soda, chloride of sodium	1·20	„
	100·00	

925. *What are the uses of the bones?*

They *protect* soft and delicate organs; they form a framework to which the organs are attached, and by which they are *kept in their places;* and they supply a *mechanism,* by which the *motions of the body are produced,* in combination with the muscles.

926. *Why is the brain placed within the skull?*

Because that delicate and vital organ, being the *centre and the root of the nervous system,* requires a position of the *greatest safety.*

927. *Why are the bones that constitute the vertebræ (back-bone) hollowed out, so as to form a continuous groove?*

Because through that groove the *spinal cord* passes out from the brain. Being in the centre of that column of bones, the spinal cord receives from them a similar protection *to that which the brain obtains from the skull.*

928. *Why is the head set upon the neck?*

Because in that position it obtains the *freest motion,* can turn in *any direction,* and is placed relatively to the other parts of the body, in that situation where it acquires *the greatest possible advantage.*

929. *Why are the eyes placed in the sockets of the skull?*

Because the bones of the skull *afford protection* to the delicate and complicated structure of the eyes, and supply points of attachment, and grooves, by which the muscles are enabled to *turn the eyes freely*, and thereby *extend the field of vision.*

930. *Why are the bones of the skull arched?*

Because in that form they acquire *greater strength*, and hence the utmost degree of safety is combined with extreme *lightness of material.*

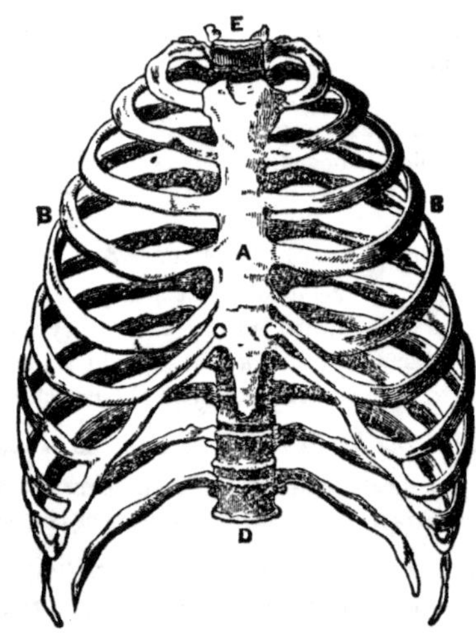

FIG. 54.—VIEW OF THE BONES OF THE THORAX, OR CHEST, SHOWING THE PROTECTION AFFORDED TO THE ORGANS OF CIRCULATION AND RESPIRATION.

A. The sternum, or breast-bone.

B B. The *ribs*, which rise a little from behind, and fall as they come forward, by which they acquire a greater flexibility.

C C. The *cartilaginous points* of the short ribs, by which their expansive and compressive powers are much increased.

D E. Part of the vertebral column, or back-bone.

931. *Why are the bones of the skull divided by sutures (seams), with points which fit into each other like small teeth?*

Because, by that arrangement, *concussions of the skull*, which might be fatal to the brain, are *deadened*, and injuries from accident *greatly modified.*

"And I will lay the sinews upon you, and will bring up flesh upon you, and cover you with skin, and put breath in you, and ye shall live; and ye shall know that I am the Lord."—Ezekiel xxxvii.

932. *Why are the heart, lungs, &c., placed within the chest?*

Because the functions of those organs require ***considerable space***, while their importance in the system of life, renders it essential that they should be ***securely protected*** from the probabilities of accident.

933. *Why are the heart and lungs enclosed for protection in a series of ribs, and not in a close case, like the brain?*

Because, by the inflation and contraction of the lungs, their *capacity is constantly changing.* When man takes a moderate inspiration, he inhales about *thirty cubic inches of air*, and the lungs increase in size *one-eighteenth of their whole capacity.* Consequently, were they enclosed in a frame of *fixed dimensions*, it must needs be, to that extent at least, larger than is necessary, when the frame is made to dilate and contract with the capacity of the lungs.

So perfect is the Almighty contrivance, that not only are the ribs made to *protect* the lungs, but, by their elasticity, and the contractions and dilations of the muscles which lie between them, they *assist the lungs in their labours*, and work with them in perfect harmony.

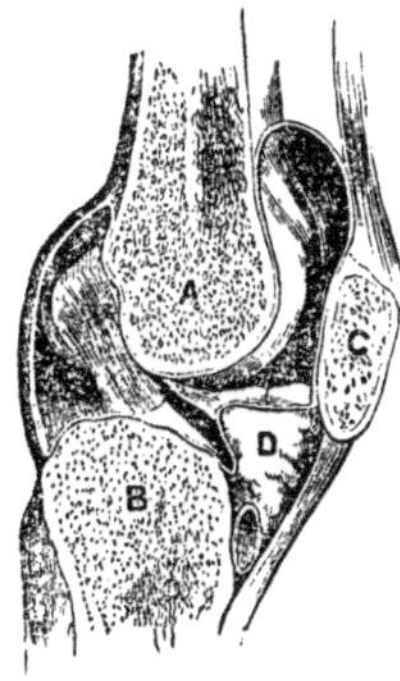

Fig. 55.—SECTION OF THE KNEE JOINT, SHOWING THE CELLULAR STRUCTURE OF BONE, BY WHICH LIGHTNESS AND STRENGTH ARE OBTAINED.

934. *Why are the bones of the arms, legs, &c., made hollow?*

Because *lightness* is thereby combined with *strength.* There is a provision by which, in the extremities of bones, where an enlarged surface is required, *lightness* is still combined with the necessary degree of strength.

The bones are made up of a *cellular formation;* and this generally occurs in parts which are much called into action, in the various movements of the body.

A. Lower part of the bone of the thigh.
B. Head of the bone of the leg.
C. The knee cap showing its relation to the other bones, and the manner in which it is enclosed by the tendons seen at Fig. 58.
D. A pad of fat, lessening the friction of the bones and modifying the shocks produced by jumping, &c.

Again he said unto me, Prophesy upon these bones, and say unto them, O ye dry bones, hear the word of the Lord."—EZEKIEL XXXVII.

935. *Why are the bones of the arms and legs formed in long shafts?*

Because a considerable *leverage* is gained, by which the advantages of *quickness of motion*, and *increase of mechanical power*, are secured.

936. *Why are the bones of the hands and feet numerous and small?*

Because the motions of the hands and feet are very *varied and complicated*. There are no less than *twenty-eight bones* in one hand and wrist; and about *as many* in a foot and ankle. To these are fastened a great number of *ligaments and muscles*, by which their varied compound movements are controlled. But for the complexity of the mechanism of our hands and feet, our motions would be extremely awkward, and many of the valuable mechanical inventions which now benefit mankind, could never have been introduced. The bones of the hands and feet are in number equal to *one-half of the whole of the bones of the body.*

CHAPTER XLVI.

937. *What are ligaments?*

Ligaments consist of bands and cords of *a tough, fibrous, and smooth substance*, by which the bones are bound together and held in their places, allowing them freedom to move, and supplying smooth surfaces over which they glide.

938. *Why are the joints bound with ligaments?*

Because the bones would otherwise be constantly liable to *slip from their places*

939. *What are tendons?*

Tendons are *long cords*, of a substance similar in its nature to ***cartilage***, by which ***the muscles are attached to the bones.***

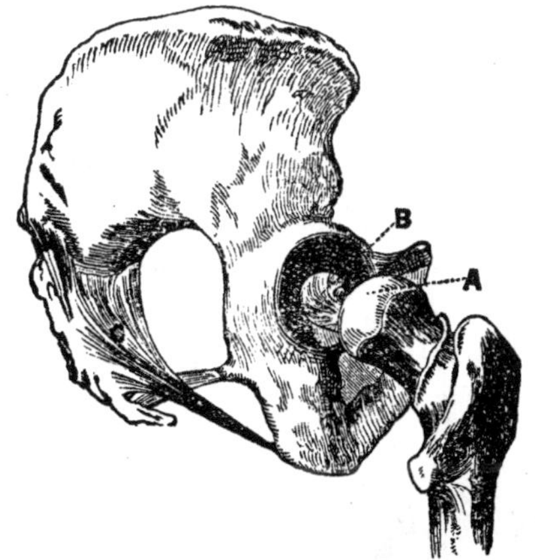

Fig. 56.—SHOWING A BALL AND SOCKET JOINT, AND THE MANNER IN WHICH LIGAMENTS ARE EMPLOYED TO HOLD BONES IN THEIR POSITIONS.

A. The *ball*, or *head* of the thigh bone.
B. The *socket*, showing the ligament in the socket, which holds the head of the bone in its place, but allows it free motion.
C. *Ligaments* tied from bone to bone, giving firmness to the parts.

940. *Why are tendons used to attach the muscles to the bones?*

Because, by this arrangement, the large muscles by which the extremities are moved, *may be placed at some distance* from the bones upon which they act, and thus the extremities, instead of being large and clumsy, are *small* and *neat*.

941. *How many muscles are there in the human body?*

There are about *four hundred and forty-six muscles* that have been dissected and described, and the actions of which are perfectly understood. But there is probably a much larger number of muscles, and of compound actions of muscles, than the skill of man has been able to recognise.

"All flesh is not the same flesh: but there is one kind of flesh of men, another flesh of beasts, another of fishes, and another of birds."—CORINTHIANS XVI.

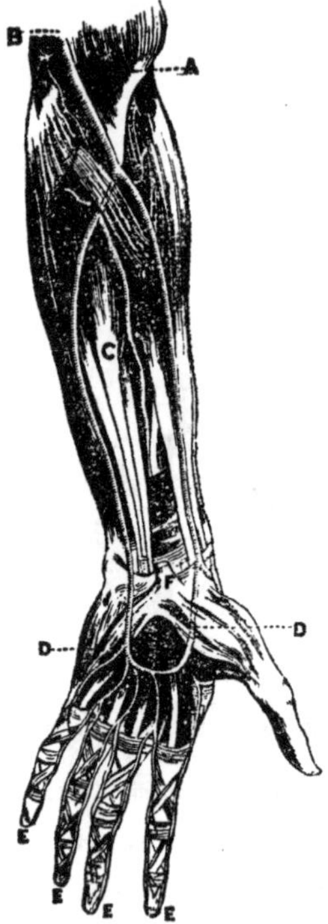

Fig. 57.—ILLUSTRATION OF THE RELATION OF MUSCLES, TENDONS, AND BONES.

942. *What is the constitution of a muscle?*

Every muscle is made up of a number of *parallel fleshy fibres*, or threads, which are bound together by a smooth and soft tissue, forming a sheath or case to the muscle, and enabling it to *glide freely* over the surfaces upon which it moves.

A. Lower extremity of the muscle which draws the fore-arm towards the upper-arm, bends the elbow, raises the hand to the head, and is powerfully exerted in pulling, lifting, &c.

C. A muscle which gives off four long *tendons*, which pass under the *ligaments* of the wrist, one to each finger, and by which the fingers are bent upon the palm of the hand, as in grasping, &c.

F. *Tendon* of a muscle which draws the little finger and the thumb towards each other.

The *ligaments* may be seen enfolding the finger-joints, and also crossing the wrist, underneath the *tendons*.

The muscles are compressed into *tendinous cords* at their ends, by which they are *united to the bones.*

They are arranged *in pairs*, having reciprocal actions—each muscle having *a companion muscle* by which the part which it moves is restored to its original position, when the influence of the first muscle is withdrawn, and the stimulus given to bring back the part.

943. *Why can we raise our fingers?*

Because muscles which lie *on the fore-arm*, and have their

"Thou hast clothed me with skin and flesh, and hast fenced me with bones and sinews."—JOB XI.

tendons fastened at the ends of the fingers, *contract*, and by becoming shorter, *draw the fingers upward*, and towards the arm.

944. *Why can we throw back the fingers after they have been raised?*

Because the muscles at the back of the arm, ***whose tendons are attached to the back of the fingers***, contract and restore them to their former position.

945. *What degree of strength do the muscles possess?*

The degree of strength of a muscle depends upon the ***healthy condition*** of the muscle, the ***amount of stimulus*** which it receives at the time of exertion, and the manner in which ***its powers are applied***.

The great muscle of the calf of the leg has been found, when removed from a dead body, to be capable of sustaining a weight equal to ***seven times the weight of the entire body***

But the contractile power of the living muscles is very great: the thigh bone has frequently been broken by muscular contractions in fits of epilepsy. And in cases where there has been a dislocation of the thigh, the head of the thigh-bone being thrown out of its socket, (Fig. 56) it has been found necessary to employ strong ropes, attached to a wheel turned by several hands, in order to ***overcome the contraction of the excited muscles, and to enable the operator to restore the bone to its place.***

946. *What is the stimulus which sets the muscles in action?*

The muscles are excited to action by ***the nerves***, which they receive from the ***spinal cord***.

947. *Why does it require the influence of the will to set the arms in motion?*

Because the muscles which form their mechanism are ***voluntary*** muscles—that is, they are subject to the ***will of man***, and influ-

"And he took him by the right hand, and lifted him up; and immediately his feet and ancle bones received strength."—ACTS III.

enced by impulses directed to them through the nervous system *by the mind*, which is the governing power.

948. *Why does the heart beat without any effort of the will?*

Because the muscles of the heart are *involutary muscles*—that is, they are *independent of the will*, and receive a *continuous nervous stimulus* which is not *under the controul of the mind.*

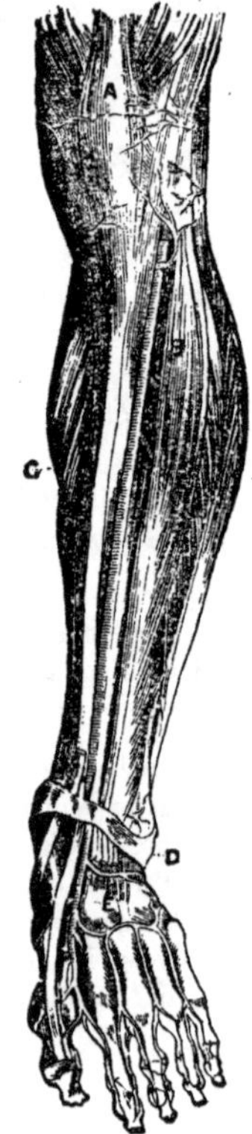

Fig. 58.—MUSCLES AND VESSELS OF THE LEG AND FOOT.

A. A large *ligament*, which covers the knee pan, or moveable bone of the knee, by which the ends of the bones of the thigh and leg are kept from slipping over each other.

B. A muscle which passes underneath the cartilages of the ankle, and gives off four *tendons*, which are distributed to the toes, and by which they are extended in elongating the foot, walking, &c.

C. Part of the muscle which forms the fleshy bulb of the calf of the leg, and which terminates in the large *tendon* attached to the heel, called the *tendon of Achilles.*

D. One of the ligaments which bind the tendons and the bones of the ankle.

E. *Arteries* proceeding from the large vessel descending the leg, by which the toes are supplied.

949. *Why are the muscles of the arms, &c., made subject to the controul of the will?*

Because, as they supply the mechanism through which we adapt ourselves to our varying wants and circumstances, it was necessary that they should be placed under the controul of the mental power, and be moved *only in accordance with man's necessities.*

"If thou sayest, Behold, we knew it not; doth not he that pondereth the heart consider it? and he that keepeth thy soul, doth not he know it? and shall not he render to every man according to his works?"—PROVERBS XXIV.

950. *Why are the motions of the heart, &c., made independent of the will?*

Because, as the necessity for the heart's motion is ***fixed and unalterable,*** the constant motion of the heart could be best secured by giving it a *fixed nervous influence,* by which it might be unfailingly prompted to fulfil its functions.

If the movements of man's heart were ***subject to his will,*** he would be constantly required to regard the operations of that organ; and so large an amount of mental care and physical exertion would have to be employed in that direction, that man's sole work would be to keep himself alive. Hence we see the goodness of the Creator in *giving* life to man, and in ***keeping the vital impulses under his divine care.***

CHAPTER XLVII.

951. ***What are nerves?***

The nerves are branches of the ***brain*** and the ***spinal cord;*** they are distributed in great numbers to all the active and sensitive parts of the body.

952. *What is the spinal cord?*

The spinal cord is a long and large cord of nervous matter, which extends from the brain through a continuous tube formed by corresponding hollows in the bones of the back. It serves as a nervous trunk *for the distribution of nerves,* just as the aorta distributes branches of blood-vessels.

953. *Why is the spinal cord placed in the grooves formed by the back-bone?*

Being a very vital part of the system, and from the delicacy of its structure liable to injuries, it is set in the back-bone for *protection;* and so great is its security that it is only by force of an unusual kind that it can be injured.

"A sound heart is the life of the flesh: but envy is the rottenness of the bones."—PROVERBS XIV.

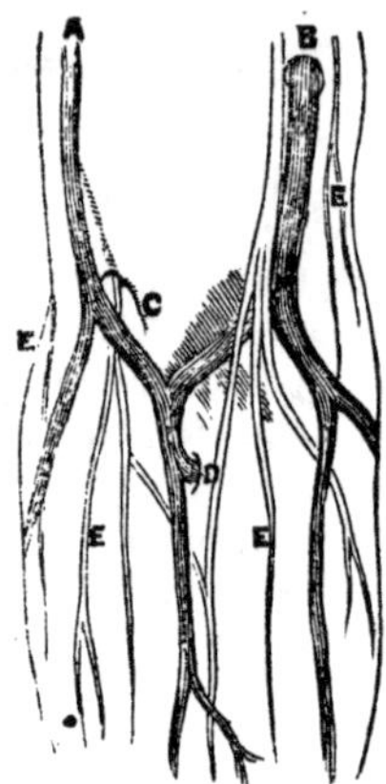

Fig. 59.—SHOWING THE DISTRIBUTION OF NERVES AND VEINS, AND ILLUSTRATING THE MANNER IN WHICH THEY PASS THROUGH THE FLESH TO REACH THE PARTS TO WHICH THEIR FUNCTIONS BELONG.

954. *How can branches proceed from it, if it is so securely encased in bone?*

Because in the bones, on each side of of the spinal cord, there are ***smaller grooves*** for the transmission of the nervous branches.

955. *Of what does the nervous system consist?*

Of the ***brain***, the ***spinal cord***, and the branches which are called ***nerves***.

A B. ***Veins*** of the fore-arm.

B. Canal formed in the muscle, through which a ***trunk-vein*** emerges.

C. Canal formed in the muscle, through which a large ***nerve*** emerges.

D. Canal through which a ***vein*** enters to communicate with the deep muscles of the arm.

956. *What is the constitution of a nerve?*

It consists of a thin membrane, or sheath, surrounding ***a greyish oily matter***, which forms the nervous marrow. In the centre of this marrow is usually found ***a small fibre***, which is supposed to be the essential part of the nerve; and most nerves consist of a number of these sheaths enclosing fibres running in parallel directions.

957. *What is the nervous fluid?*

The term ***nervous fluid*** is used to express our ideas of the mode by which the brain and spinal cord influence the remote parts: just as we say the *electric fluid*, without knowing that such a fluid exists. It is the most convenient form of expression.

958. *How many classes of nerves are there?*

There are :—

1. The nerves of ***motion***.
2. The nerves of ***sensation***.

"Having many things to write unto you, I would not write with paper and ink but I trust to come unto you, and speak face to face, that our joy may be full."—II JOHN.

3. The nerves of *special sense.*
4. The nerves of *sympathy.*

959. *What are the nerves of motion?*

The *nerves of motion* are those which, in obedience to the will, *stimulate the muscles to act,* and apportion the amount of stimulation they convey to the degree of exertion required.

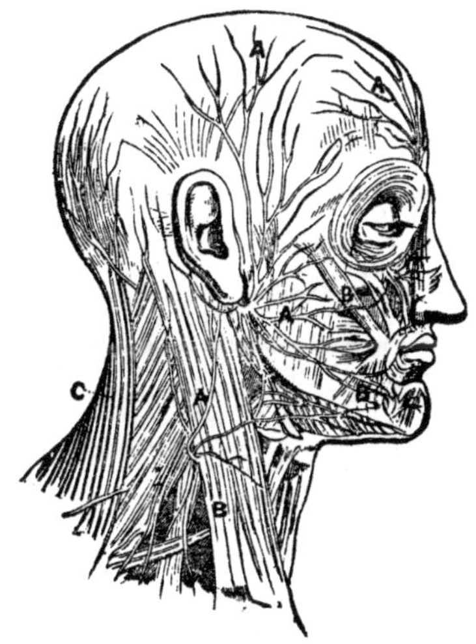

Fig. 60.—MUSCLES OF THE HEAD AND FACE, WITH NERVES DISTRIBUTED THERETO.

A A A. The *facial nerve* emerging from underneath the ear, and distributing branches to the cheeks, temple, forehead, &c. This nerve excites the muscles of the face, and is chiefly instrumental in producing the expressions of the countenance under the changing emotions of the mind.

B B B. *Muscles* by which various motions are imparted to the head, face, mouth, &c., under the *stimulus of the nerves.*

960. *What are the nerves of sensation?*

The *nerves of sensation* are those which *impart a consciousness to the brain* that its commands to the nerves of motion have been obeyed, and how far they have been fulfilled.

"Oh that men would praise the Lord for his goodness, and for his wonderful works to the children of men."—PSALM CVII.

Let us perform a simple experiment, which will more clearly illustrate the phenomena of *motion* and of *sensation*, which we are now describing, than a great deal of writing upon the subject. You hold in your hand this book: close it, and set it upon the table; lay your hands passively upon your lap, and then *will* your hand, to take up the book, which is the same as to say, *command* your hand to take up the book. What occurs? The hand, immediately obeying your desire, stretches forward to the book, and takes hold of it. How do you know that you have hold of it? You *see* that you have: but were your eyes closed, you would be equally aware that the hand had reached the book, and fulfilled your wishes. It is by the nerves of *sensation* that you are made aware that the hand has fulfilled your instructions.

Consider what took place in the simple action. In the first instance, a desire arose in your mind to take up the book. The *brain* is the organ of the mind; and having branches either proceeding from itself, or from the spinal cord, to every part of the body—branches that traverse like telegraphic wires throughout every part of the system,—it transmitted instructions along the nerves that proceed to the muscles of the arm and hand, directing them to take up the book. This was done instantly; and as soon as it was done you became conscious that your will had been obeyed—because *the nerves sent back a sensation to the brain acquainting it that the book had been taken up*, and that at the moment of the dispatch it was in the firm hold of the hand.

In all the varied motions of the body this double action of the nerves takes place. It is obvious that without an *outward* impulse from the brain, upon which the desire of the mind first made an impression, no motion of the muscles of the arm and the hand could have taken place; and it is also obvious that without an *inward* impulse from the nerves to the brain you would not have known that the muscles had fulfilled your instructions. The hand might have dropped by the side of the book, or have gone too far, or not far enough, and you would not have been waare of the result, but for an inward communication through the nerves.

We are not now speaking of the nerves which endow us with the sense of *feeling;* because they are regarded as separate and distinct from those nerves that produce in us consciousness of muscular response. When we walk, rise, or sit, we are made conscious, without any special feeling being exerted, that the muscles have placed the limb, or the body, in the desired position, that it is set down safely and firmly, and that we may repose upon it securely without further attention. We refer the impressions made by the book upon the nerves of the hand, and which enable us to tell whether it feels hot or cold, whether its surface is rough or smooth, and so on, to the special sense of *feeling*. The consciousness of muscular action is a separate and distinct function; and it is generally believed that the same nerves that convey the command of the will outward, bringing back the intimation that the will has been obeyed, but that *different fibres* of the nerves convey the *outward* and the *inward* impulses. A single nerve may therefore be likened to a *double wire* connected with the electric telegraph: one transmitting despatches in one direction, and the other in the opposite direction.

961. *What are the nerves of special sense?*

The nerves of special sense are those through which we *hear, see, feel, smell,* and *taste*.

962. *What are the nerves of sympathy?*

The nerves of sympathy, or the system of ***sympathetic nerves***, are those which are distributed to the *internal organs*, and which are independent of the will. They regulate the motions of the heart, the lungs, the stomach, &c., and stimulate the organs of secretion, so that those organs *work in harmony with each other*.

As the internal organs are all more or less dependent upon each other, and unite their functions for similar ends, it is obvious that there should prevail among them a *mutual consciousness* of their state. Otherwise, when the stomach had formed chyme, the liver might have no bile ready to fulfil its office; the absorbents might be in a state of rest at the moment when nutrition was set before them; and the heart might beat slowly, while the lungs were in active exertion to obtain additional blood to support an active exercise. The sympathetic system of nerves therefore ***regulates and harmonises these internal functions.***

CHAPTER XLVIII.

963. *Why do we see objects?*

Because the light which is reflected from them enters our eyes and produces images of their forms upon a membrane of nerves called the *retina*, just as images are produced upon a mirror.

964. *Why does this enable us to see?*

Because the membrane which receives the images of objects is connected with the *optic nerve* which transmits to the brain impressions made by the reflections of light, just as other nerves convey the effects of feeling, hearing, tasting, &c.

965. *Why are we enabled to move our eyes?*

Because various muscles are so placed in relation to the eyeball, that their contraction draws the eye in the direction required. We are thus enabled to adjust the direction of the eye to the position of the objects we desire to see, in other words to ***set the mirror in such a position that it will receive the reflection.*** (See 517.)

"Truly the light is sweet, and a pleasant thing it is for the eyes to behold the sun."—ECCLESIASTES XI.

966. *Why are we enabled to see large objects upon so small a surface?*

Because the lenses and humours of the eye *collect the rays of light* coming from every direction, and, *bringing them into a focus,* transmit them to the retina, where each ray impresses upon the nervous surface the qualities it received from the object which reflected it.

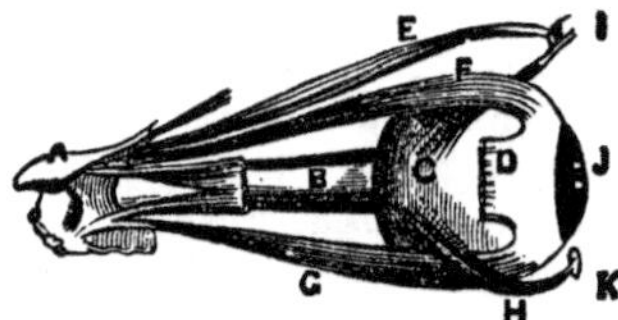

Fig. 61.—THE EYEBALL AND ITS MUSCLES.

A. Portion of *bone* through which the optic nerve passes in its communication between the brain and the eye.

B. The *optic nerve,* from before which an *external muscle* has been cut away, leaving its two attachments.

C. The *globe* of the eye.

D. The muscle which turns the eye *outward,* and which is counteracted by a muscle on the other side.

E. The muscle which passes through a loop, or staple of cartilage I, and *turns the eye obliquely.* It is counteracted by a muscle situated underneath.

F. The muscle situated underneath, which *turns the eyeball upwards,* and is counteracted by

G. The muscle which *turns the eyeball downwards.*

H. The muscle attached to a bone which *turns the eyeball upwards.*

I. The *cartilaginous loop* through which a muscle passes.

J. The front chamber of the eye filled with a clear fluid.

K. Fragment of the bone by which one of the muscles is fastened.

967. *Why do some persons squint?*

Because it sometimes happens that a muscle of the eye *acts too powerfully* for its companion muscle, and draws the eye too much on one side.

968. *Why does the pupil of the eye look black?*

Because the pupil is an *opening* through which the rays of light pass into the chamber of the eye. There is, therefore, nothing in the pupil, of the eye to reflect light.

"Keep me as the apple of thine eye; hide me under the shadow of thy wings."—PSALM XVII.

969. *Why is the pupil of the eye larger sometimes than at others?*

Because the *iris*, a ring of extremely fine muscles which surround the pupil, contracts when too much light falls upon the retina, and dilates when the light is feeble. It therefore enlarges or diminishes the size of the pupil to *regulate the admission of light.*

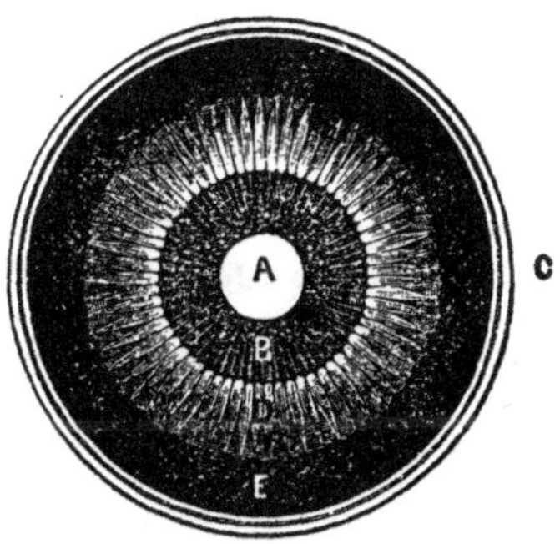

Fig. 62.—SECTION OF THE EYE SEEN FROM BEHIND

A. The *pupil* of the eye through which the light enters.

B. The *iris*, which dilates or contracts, and thereby increases or lessens the size of the *pupil.*

C C. The three coats of the eye, called the *sclerotic*, *choroid*, and *retina.*

D. The *ciliary processes*, or hair-like muscles, which have a slight vibratory motion which they impart to the fluids of the eye.

E. The dark coat of the *choroid*, the coat forming the *retina* removed.

970. *Why have we two eyes?*

Because the field of vision is thereby *much extended;* the *intensity* of sight is also increased, the impressions upon the brain being clearer and better defined, just as in a *stereoscope* the effect of vision is heightened by a double picture; the sense of sight being more *constantly* exercised than any other sense during our waking moments, *one eye is frequently called upon to give rest to the other;* and the important faculty of vision, being endangered by the necessary exposure of some parts of the eye, and the equally necessary delicacy of an organ formed to receive impressions from so ethereal an element as light, is rendered the more secure to us, since though one eye may become enfeebled, diseased, or wholly lost, *the other eye will retain the blessing of sight.*

971. *Why, having two eyes, and each eye receiving a reflection upon its retina, does the brain experience only one impression of an object?*

Because, besides those optical laws which bring upon the two retinas the exactly corresponding images of the same objects, the optic nerves *meet* before they reach the brain, *and blend the impulses which they convey.*

972. *Why are the eyes provided with eye-lids?*

Because the eyes require to be *defended* from floating particles in in the air, and to be kept *moist and clean.* The eye-lids form the shutters of the eye, defending it when waking, by closing upon its surface whenever danger is apprehended, moistening its surface when it becomes dry, and covering it securely during the hours of sleep.

973. *Why are the eye-lids fringed with eye-lashes?*

Because the eye-lashes assist to modify the light, and to protect the eye, without actually closing the eye-lids. When the eye-lids are partially closed, as in very sunny or dusty weather, the eye-lashes cross each other, forming a kind of shady lattice-work, from the interspaces of which the eye looks out with advantage, and sees sufficiently for the guidance of the body.

974. *Why are we able to see at long or short distances?*

Because the *crystalline lens* of the eye is a moveable body, and is pushed forward, or drawn back by fine muscular fibres, acccording to the distances of the objects upon which we look. By these means its *focus* becomes adjusted.

975. *Why do we wink?*

Because, by the repeated action of winking, *the eye is kept moist and clean,* and the watery fluid secreted by little glands in the eye-lids, and at the sides of the eye, is spread equally over the surface, instead of being allowed to accumulate. But the action of

"And the eye cannot say unto the hand, I have no need of thee; nor again the head to the feet, I have no need of you."—CORINTH. XII.

winking, or brightening the eye, is so instantaneous that it does not impede the sight.

976. *Whence are the humours and secretions of the eye derived?*

From the blood, which flows abundantly to the eyes, and is circulated in capillary vessels that are spread out upon the membranous coats of the eye-balls.

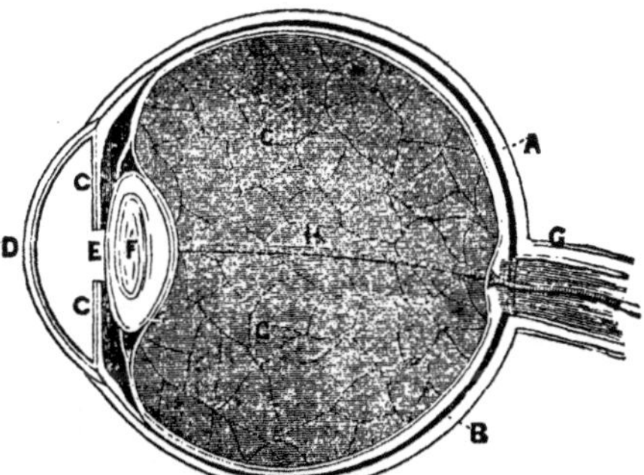

Fig. 63.—SECTION OF THE EYE.

A and B. The *sclerotic*, *choroid*, and *retina*, the three layers or coats which form the walls of the globe of the eye, and enclose its humours.
C C. The *iris*.
D. The front chamber of the eye, filled with watery humour.
E. The *pupil*, through which the rays of light pass to
F. The *crystalline lens*.
G G. The *vitreous humour* enclosed in cells formed by the *hyaloid membrane*.
H. An *artery* which supplies blood to the *crystalline lens*, and which passes through the centre of the *optic nerve*.
G. The *optic nerve*, showing the sheath in which the nerve is enclosed.

977. *Why do tears form in the eyes?*

Because, under the emotions of the mind, the circulation of blood in the brain, and in its nearest branches, becomes considerably quickened. The eyes receive a larger amount of blood, and the secretion of the lachrymal glands being increased, the fluid overflows, and tears are formed. The use of tears is probably *to keep the eyes cool during the excitement of the brain.* They are formed also during *laughing*, but less frequently.

"If the whole body were an eye, where were hearing? if the whole were hearing where were smelling?"—CORINTHIANS XII.

978. *Why do we feel inconvenienced by sudden light?*

Because an excess of light enters the eye before the *iris* has had *time to adjust the pupil* to the amount of light to be received.

979. *Why if we look upon a very bright light, and then turn away, are we unable to see?*

Because the *iris* has so reduced the pupil while we were looking at the bright light, that immediately upon turning to a darker object, *the pupil is too small* to admit sufficient rays to enable us to see.

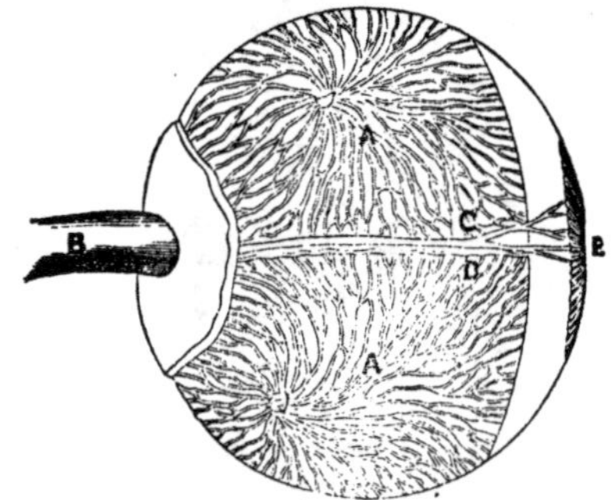

Fig. 64.—CAPILLARY BLOOD-VESSELS OF THE EYE.

A A. Capillary veins distributed over the *sclerotic coat.*
B. One of the trunks of the *optic nerve.*
C. A *nerve* communicating with the *ciliary processes.*
D. A *vein* running parallel with the nerve to the *ciliary processes.*
E. Side view of the *iris.*

980. *Why do we see better after a short time?*

Because the *iris* has relaxed and enlarged the pupil, therefore *we receive more rays of light* from the comparatively dark object, and are enabled to see it more clearly.

981. *Why do cats, bats, owls, &c., see in the dark?*

Because their eyes are made highly sensitive to *small quantities of light.* It is also believed that there are certain properties of light which affect their eyes, but do not affect ours. In other words that there are some rays which are luminous to them which are not luminous to us. Hence they find *light* in what we call *darkness*.

982. *Why does the pupil of a cat's eye appear nearly closed by day?*

Because the cat's eye is so sensitive to light that the iris *closes the pupil almost entirely* to shut out the too powerful light.

CHAPTER XLIX.

983. *Why do we hear?*

Because the *tympanum* of the ear *receives impressions from sounds*, and transmits those impressions to the brain in a similar manner to that in which the retina of the eye transmits the impressions made upon it by light.

984. *Why is one part of the ear spread out externally?*

The external ear is a *natural ear-trumpet*, and serves to collect the vibrations of sound, and to conduct them towards the internal ear.

985. *Why is the ear allowed to project, whilst the eye is carefully enclosed?*

Because the external ear, being formed of tough cartilaginous substance, and being very simple in its organisation, is but little liable to injury.

986. *Why do hairs grow across the entrance of the ears?*

Because they prevent the intrusion of insects, and of particles of dust, by which otherwise the faculty of hearing would be impaired.

The insect called the *ear-wig* is popularly supposed to be so named from its tendency to get into the human ear, and cause pain and madness by penetrating to the brain. An earwig, however, is no more likely to get into the ear than any other insect whose habit it is to penetrate the corollas of flowers; and should an insect enter the ear, it could get no further than the *membrane of the tympanum*, which spreads all over the auditory passage, just as the parchment of a drum spreads over the entire circumference of that instrument. The fact is, that the wing of the insect, when spread, *resembles the external ear* in shape. It is similar to the wing of the stag beetle (*see* illustration), and this fancied resemblance of the wing of the insect to the ear of man may have given rise to the name of *ear wing*, which became corrupted to *ear-wig*

"Doth not the ear try words? and the mouth taste his meat."—JOB XII.

987. *Why is wax secreted at the entrance of the ear?*

Because, by the peculiar resinous property which it possesses, *it improves the sound-conducting power* of the auditory canal through which it prevails.

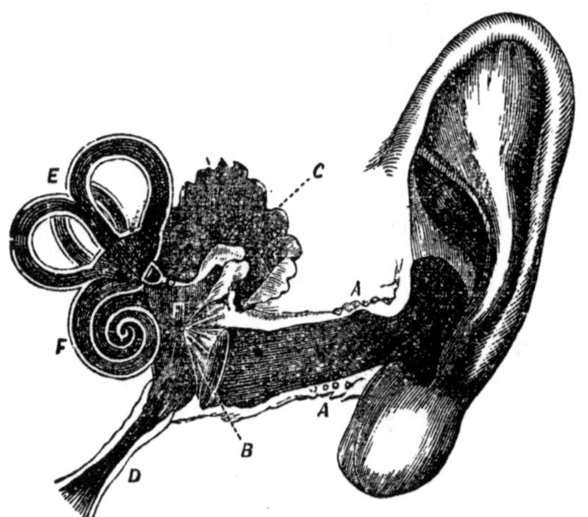

Fig. 65.—THE STRUCTURE OF THE EAR.

A A. Glands which secrete wax in the walls of the tube of the ear.

B. The membrane of the *tympanum*, or drum of the ear, formed in the shape of a funnel.

C C. Bones which act as a sort of sounding-board to the ear, giving strength to the vibrations.

D. The Eustachian tube, which opens into the root of the mouth, and which serves to preserve an equilibrium in the density of the air occupying the tubes of the ear.

E and F. The *labrynth* of the ear, consisting of folds of membraneous tubes, filled with fluid, which serves to undulate with the vibrations of the *tympanum*, and thus gives clearness and precision to the sounds.

The *auditory nerves* are distributed in the tubes above described (the *vestibule* and the *cochlea* E F), and the nerves receive their impressions from the undulations of the fluid.

988. *Why do we sometimes hear singing noises in the ear?*

Because the ear is liable to inflammation from various causes, and

"Apply thine heart unto instruction, and thine ears to the words of knowledge."—PROVERBS XXIII.

when the blood flows unduly through the vessels of the ear it *produces a slight sound.*

989. *Why do people become deaf?*

Because the ear may be injured in various ways: the tympanum may be impaired, the fluid of the ear dried up, or the nerves be pressed upon by swellings in the surrounding parts. When, therefore, the *mechanism of hearing* is impaired, the sense of hearing becomes weakened, or altogether lost.

990. *Why do persons accustomed to loud noises feel no inconvenience from them?*

Because the *sensitiveness* of the nerves of the ear becomes deadened. They do not convey to the brain such intense impulses when they are frequently acted upon by loud sounds.

991. *Why do persons engaged in battle often lose their hearing?*

Because the vibrations caused by the sounds of artillery are so violent that they overpower the mechanism of the ear, and frequently *rupture the connection of the fine nervous filaments* with the textures through which they spread.

The violent concussions of the air produced by volleys of cannon, or by loud peals of thunder, have an overpowering effect upon persons nervously constituted, and upon the organ of hearing, which is more especially affected. As persons have been struck blind by intense light, so others have been deafened by intense sounds. In 1697 a butcher's dog was killed by the noise of the firing to celebrate the proclamation of peace. Two troops of horse were dismounted, and drawn up in a line to fire volleys. At the moment of the first volley a large and courageous mastiff, belonging to a butcher, was lying asleep before the fire. At the noise of the first volley the dog started up, and ran into another room, where it hid itself behind a bed; on the firing of the second volley, it ran several times bout the room, trembling violently; and when the hird volley was fired it ran around once or twice with great voilence, and then dropped down dead, with blood flowing from its mouth and nose. Persons who are painfully affected by loud noises should put a little wool in their ears when such noises are occurring; they will thereby save themselves from temporary inconvenience, and probably preserve the sense of hearing from permanent injury.

992. *Why do we smell?*

Because minute particles of matter, diffused in the air, come in

"And the Lord God formed man of the dust of the ground, and breathed into his nostrils the breath of life; and man became a living soul."—GENESIS II.

contact with the filaments of the *olfactory nerve*, which are spread out upon the walls of the nostrils, and those nerves transmit impressions to the brain, constituting what we call the *odour of substances*.

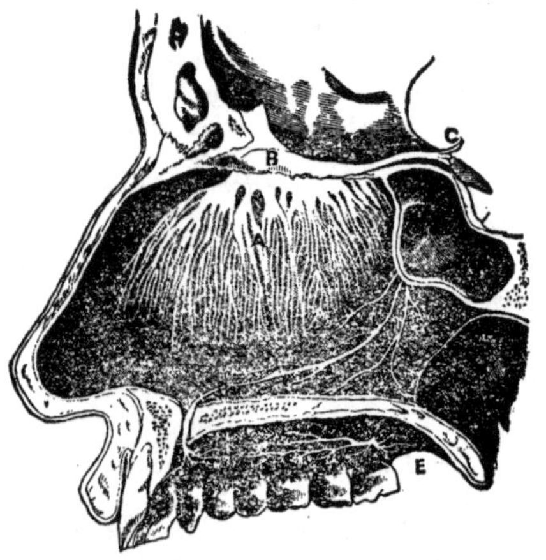

Fig. 66.—SHOWING THE DISTRIBUTION OF THE NERVOUS FILAMENTS UPON SENSITIVE MEMBRANES.

A. The *olfactory nerve*, distributed in minute branches upon the membrane of the nostril.

B. The *bulb* of the *olfactory* nerve.

C. The *roots* from which the olfactory nerve originates.

D E. *Nerves of the palate*, showing the manner in which they are passed through the bones of the roof of the mouth.

993. *Why do hairs grow across the passages of the nostrils?*

Because they form a *defence* against the admission of dust and insects, which would otherwise frequently irritate the nervous structure of the nose.

994. *Why are the nostrils directed downwards?*

Because, as odours and effluvia *ascend*, the nose is directed towards them, and thereby receives the readiest intimation of those bodies floating in the air which may be pleasurable to the sense, or offensive to the smell, and injurious to life.

995. *Why is the nose placed over and near the mouth?*

Because, as one of the chief duties of that organ is to *exercise a watchfulness* over the purity of the substances we eat and drink, it is placed in that position which enables it to discharge that duty with the greatest readiness.

CHAPTER L.

996. *Why do we taste?*

Because the tongue is endowed with *gustatory* nerves, having the function of *taste* as their *special sense,* just as the *optic,* the *auditory,* and the *olfactory* nerves, have their special duties in the eyes, ears, and nose.

997. *Why do some substances taste sweet, others sour, others salt, &c.?*

It is believed that the impressions of taste arise from the various *forms of the atoms of matter* presented to the nerves of the tongue.

998. *Why do we taste substances most satisfactorily after they have remained a little while in the mouth?*

Because the nerves of taste are most abundantly distributed to the under surface of the tongue; and when solid substances have been in the mouth a little while, they impregnate the saliva of the mouth with their particles *and come in contact in a fluid solution with the gustatory nerves.*

999. *Why if we put a nub of sugar to the tip of the tongue has it no taste?*

Because the gustatory nerves are *not distributed to that part of the tongue.*

1000. *Why, when we draw the tongue in, do we recognise the sweetness of the sugar?*

Because the dissolved particles of sugar are *brought in contact* with the nerves of taste.

1001. *Through what nerves are we made sensible of the contact of sugar with the tip of the tongue?*

Through the nerves of *feeling*, which are abundantly distributed to the tongue to guide it in its controul over the mastication of food.

1002. *Why do conniseurs of wines close their mouths and distend their chins for a few seconds, when tasting wines?*

Because they thereby bring the wine in contact with the under surface of the tongue, *in which the gustatory nerves chiefly reside.*

1003. *Why do they also pass the fumes of the wines through their nostrils?*

Because *flavour*, in its fullest sense, comprehends not only the *taste*, but the *odour* of a substance; and, therefore, persons of experience attend to both requisites.

The various conditions of taste are defined to be:—

1. Where sensations of *touch* are alone produced, as by glass, ice, pebbles, &c.

2. Where, in addition to being *felt* upon the tongue, the the substance excites sensation in the *olfactory nerves*, as by lead, tin, copper, &c.

3. Where, besides being *felt*, there are peculiar sensations of *taste*, expressive of the properties of bodies, as salt, sugar, tartaric acid, &c.

4. Where, besides being *felt* and *tasted*, there is an *odour* characteristic of the substance, and essential to the full developement of its flavours, as in cloves, lemon-peel, carriway-seed, and aromatic substances generally.

1004. *Why do we feel?*

Because there are distributed to various parts of the body fine nervous filaments, which have for their special duty the transmission to the brain of impressions made upon them *by contact* **with substances.**

"The works of the Lord are great, sought out of all them that have pleasure therein."—PSALM CXI.

1005. *In what parts of the body does the sense of touch more especially reside?*

In the points of the fingers and in the tongue. By laying a piece of paper upon a table, and upon the paper a piece of cloth, on the piece of cloth a bit of silk, and on the bit of silk a piece of leather, so that the edge of each would be exposed to the extent of half-an-inch, it would be possible by the touch to tell when the finger passed successively over the leather, silk, cloth, or paper, and arrived on the table.

Those impressions of touch must have been communicated, with their extremely nice distinctions, to the sensitive nerves that lie underneath the skin, and must have been transmitted all the way through the arm to the brain, although the touch itself was so light as scarcely to be appreciable with regard to the force applied.

A hair lying on the tongue will be plainly perceptible to the touch of the tongue; and the surface of a broken tooth will often causes the tongue great annoyance, by the acute perception it imparts of the roughness of its surface.

The toes are also highly sensitive, though their powers of touch are seldom fully developed. Persons who have lost their arms, however, have brought their feet to be almost as sensitive as fingers. Blind persons increase, by constant exercise, their powers of touch to such a degree that they are able to read freely by passing their fingers over embossed printing; and they have been known to distinguish *colours* by differences in their grain, quite unappreciable by other persons.

1006. *Why is feeling impaired when the hands are cold?*

Because, as the blood flows slowly to the nerves, they are less capable of that perception of touch which is their *special sense.* The skin contracts upon the nervous filaments, and *impairs the contact* between them and the bodies which they touch.

1007. *Why do the fingers prick and sting when they again become warm?*

Because, as the warmth expands the cuticle, and the blood begins to flow more freely through the vessels, *the nerves are made conscious of the movements of the blood,* and continue to be so until the circulation is equally restored to all the parts.

1008. *Why do persons whose legs and arms have been amputated fancy they feel the toes or fingers of the amputated limb?*

Because the nervous trunk which formerly conveyed impressions from those extremities remains in the part of the limb attached to the body. *The mind has been accustomed to refer the impulses received through that nervous trunk to the extremity where the sensations arose.* And now that the nerve has been cut, the painful sensation caused thereby is referred to the extremity which the nerve supplied, and the sufferers for a time appear to *continue to feel the part which they have lost.*

CHAPTER LI.

1009. *Why do we perspire?*

Because the skin is filled with very *minute pores*, which act as outlets for a portion of the water of the blood, that serves to *moisten and cool* the surface of the body, and to carry away some of the matter *no longer needed in the system.*

1010. *How is the perspiration formed?*

By very small *glands*, which lie embedded in the skin. It is estimated that there are about 2,700,000 perspiratory glands distributed over the surface of the body, and that these glands find outlets for their secretion through no less than *seven millions of pores.*

1011. *What is insensible perspiration?*

Insensible perspiration is that *transmission of watery particles through the skin* which is constantly going on, but which takes place so gently that it cannot be perceived. It is, however, very important in its results, as no less than *from twenty to thirty-three ounces of water may pass imperceptibly through the skin in twenty-four hours.*

1012. *What is sensible perspiration?*

"And Elisha sent a message unto him, saying, Go and wash in Jordan seven times, and thy flesh shall come again to thee, and thou shalt be clean."—II KINGS V.

Sensible perspiration is that moisture which exudes upon the skin *in drops large enough to be perceptible,* when the body is heated by exercise or other means.

1013. *Why does a sudden change from heat to cold bring on illness?*

Because the effect of cold *arrests the action of the vessels of the skin,* and suddenly throws upon the internal organs the excretory labour which the skin should have sustained.

1014. *Why does a chill upon the skin frequently produce inflammation of the lungs?*

Because the lungs and the skin together discharge the chief proportion of the watery fluid of the body. *When the skin's action is checked, the lungs have to throw off a much greater amount of fluid.* The lungs, therefore, become *over worked,* and inflammatory action sets in.

1015. *Why does cleanliness promote health?*

Because every atom of dirt which lodges upon the surface of the body serves to clog and check the working of those *minute pores,* by which much of the fluid of the body is changed and purified.

In the internal parts of the system, the Creator has made ample provision for cleanliness. Every organ is so constituted that it cleanses and lubricates itself. Every surface of the inner body is perfectly clean, and as soft as silk

Nature leaves to man the care of those surfaces which are under his immediate observation and controul; and he who, from idleness, or indifference to nature's laws, is guilty of personal neglect, *opposes the evident intentions of the Creator,* and must sooner or later pay the penalty of disobedience.

1016. *Why does exercise promote health?*

Because it *assists all the functions upon which life depend.* It quickens the circulation, and thereby nourishes every part of the body, causing the bones to become firm, and the muscles to become full and healthy. It promotes breathing, by which oxygen is taken into the system, and carbon thrown off, and thereby it produces a higher degree of organic life and strength than would otherwise exist. It promotes perspiration, by which, through the millions of pores of the skin, much of the fluid of the body is changed and purified. And it induces that genial and diffused warmth, which is one of the chief conditions of a high degree of vitality.

1017. *Why do we feel fatigue?*

Because those organs which stimulate the mechanism of the body to act, *themselves require rest and repair.* When the brain and nerves arrive at that state, they make their condition known to the system generally, by indications which we denominate *fatigue.*

1018. *Why, after rest, do we return invigorated to our labours?*

Because the nervous system has accumulated, during the hours of rest, a fresh amount of that *vital force* which we call the nervous fluid, and by which the various organs of the body are excited to perform the duties assigned to them.

1019. *What is sleep?*

Sleep is understood to be that state of the body in which *the relation of the brain to some parts of the body is temporarily suspended.*

There are some parts of the body that *never sleep:* such are the heart, the lungs, the organs of circulation, and those parts of the nervous system that direct their operations.

But when sleep overtakes the system, it seems as if the *relations* of those parts under the controul of the will were temporarily suspended; as if, for instance, those nerves which move the arms, the legs, the eyes, the tongue, &c., were all at once unfastened, just as the strings of an instrument are relaxed by the turning of a key, or the throwing down of a bridge over which they were stretched.

What is meant by the temporary suspension of the relation of the brain to some parts of the body, may be thus explained. Notice a man when he sits dosing in a chair: at first his head is held up, the brain controlling the muscles of the neck, and keeping the head erect. But drowsiness comes on, the brain begins to withdraw its influence, and the muscles of the neck becoming as it were "unstrung," the head drops down upon the breast. But the sleep is unsound, and disturbed by surrounding noises. The brain is therefore frequently excited to return its influence to the muscles, and draw up the head of

"Yet a little sleep, a little slumber, a little folding of the hands to sleep: So shall thy poverty come as one that travelleth; and thy want as an armed man."—PROVERBS XXV.

the sleeper. He gives a sudden start, every muscle is tightened in an instant, up goes the head, the eyes open, the ears listen, until a feeling of security and composure returns; the sleep again deepens, *the nervous connection is again withdrawn*, and then down drops the head as before.

1020. *Why do we dream?*

Dreams appear to arise *from the excitement of the brain during those hours when its connection with the other parts of the living organism is suspended.* For instance: a man dreams that he is pursued by a furious animal, and the mind passes through all the excitement of flying from danger; but the *connection* between the moving power, and the machinery of motion being suspended, no motion takes place. The same impressions upon the brain, when the nerves were "strung" to the muscles, would have caused a rapid flight, and a vigorous effort to escape from the apprehended danger.

1021. *Why do suppers, when indigestible substances are eaten, produce dreaming?*

Probably because, as the digestive organs are oppressed, and those parts of the nervous system which stimulate the organs of digestion are *excited by excessive action*, those portions of the brain which are not immediately employed by the digestive process are disturbed by that *sympathy* which is observed to prevail between the relative parts and functions of the body.

1022. *Why do we yawn?*

Because, as we become weary, the nervous impulses which direct the respiratory movements are enfeebled. It has been said that those movements are involuntary, and that the parts engaged in producing them are not subject to fatigue. But the operation of breathing is, *to some extent, voluntary*, though when we cease to direct it voluntarily, it is involuntarily continued by organs which know no fatigue.

When, therefore, we feel weary—still controuling our breathing in our efforts to move or to speak—there frequently arrives a period when, for a few seconds, the respiratory process is suspended. It seems to be the point at which the voluntary nerves of respiration are about to deliver their office over to the involuntary nerves; but

"And it shall be, when they say unto thee, Wherefore sighest thou? that thou shalt answer, For the tidings, because it cometh; and every heart shall melt, and all hands shall be feeble."—EZEKIEL XXI.

the pause in the respiration has caused a momentary deficiency of breath, and the involuntary nerves of respiration, coming suddenly to the aid of the lungs, cause a spasmodic action of the parts involved, and *a yawn*, attended by a *deep inspiration to compensate for the cessation of breathing*, are the result.

1023. *Why do we cough?*

Because the respiratory organs are excited by the presence of some body foreign or unnatural to them. A cough is an effort on the part of the air tubes to free themselves from some source of irritation. And so important are the organs of breathing to the welfare of the body, that the muscles of the chest, back, and abdomen, *unite in the endeavour to get rid of the exciting substance.*

1024. *Why do we sneeze?*

Because particles of matter enter the nostrils and excite the nerves of feeling and of smell. In sneezing, as in coughing, the effort is to free the parts affected from the intrusion of some matters of an objectionable nature. And in this case, as in the former one, there is a very general sympathy of other organs with the part affected, and an energetic effort to get rid of the evil.

1025. *Why do we sigh?*

The action of sighing arises from very similar causes to those of yawning. But in sighing, the nervous depression is caused by *grief*; while in yawning, it is the result of *fatigue*. In sighing, the effect is generally caused by an *expiration*—in yawning by an *inspiration*. The mind, wearied and weakened by sorrow, omits for a few seconds to continue the respiratory process; and then suddenly there comes an involuntary expiration of the breath, causing a faint sound as it passes the organs of the voice.

1026. *Why do we laugh?*

Laughing is caused by the very opposite influences that produce sighing. The nervous system is highly excited by some external cause. The impression is so intense, and the mind so fixed upon it, that the respiratory process is irregular, and uncontrolled. Persons excited to a fit of laughter generally hold their breath until they can hold it no longer, and then suddenly there is a quick expiration causing eccentric sounds, the mind being too intently fixed upon the cause of excitement, *either to moderate the sounds, or to controul the breathing.*

1027. *Why do we hiccough?*

Hiccough is caused by a spasmodic twitching of the diaphragm, a thin muscular membrane which divides the chest from the abdomen. It generally arises from sympathy with the stomach; and it is highly probable that the muscular twitches and jerks are so many efforts on the part of the diaphragm to *assist the stomach to get rid of some undigested matter.*

1028. *Why do we snore?*

Snoring is caused by air sweeping through the passages that lead from the mouth through the nostrils, and which, in our waking moments, are capable of certain muscular modifications to adapt them to our breathing. But as in sleeping the nervous controul over them is withdrawn, they are left to the action of the air which, in sweeping by them, *sets them in vibration.*

We have endeavoured, by the employment of the simplest language, and by reference to some of the most familiar phenomena of nature, to impart to the reader a clear conception of those sublime laws which controul our being, and afford evidence of the goodness and power of that Almighty God to whom we are indebted for the life that we enjoy, and the varied and beautiful existences which, to the rightly constituted mind, make the earth a vast aggregation of interesting objects. We will now, before we pass on to the final section of our work, review some of the more important facts that have been communicated, and devote a few pages to meditations upon the formation of the human body—that wonderful temple of which each of us is a tenant.

We have described man's organisation. What is that organisation for? *It is to make use of the elements upon which man exists.* The lungs make use of the air; the eye makes use of the light; the stomach, and the system generally, make use of water; every part of the body uses heat; and all parts of the system demand food. The hand feeds as constantly as the mouth. The mouth is the receptacle of food, by which the body is to be fed; the stomach is the kitchen in which food is prepared for the use of the body; and the blood-vessels are the canals through which the food is sent to those members of the body that are in need of it. When we speak of man's "organs" or "members," we speak of those parts of the living machinery by which the elements are used up, or employed, for man's benefit. And this view of the subject, bearing in mind that the body is held together as the temple of a living Spirit, superior to mere flesh and blood, gives us a higher and clearer perception of the distinction between the body and the soul than that which we might otherwise entertain. The body is a machine, working for the spirit, which is its owner. While the machine works, the spirit directs and influences its actions. But when the machine stops, the spirit resigns its power over a ruined temple, quits it, and flies to a region where, as a spirit, it becomes subject to a new order of existence consistent with its severance from earthly things and laws, and there it enters upon its eternal destiny, according to the judgments and appointments of God. It is no longer dependent upon a relation between spiritual and material laws.

Suppose that the air which man breathes, instead of returning from his lungs clear and imperceptible to sight, were tinged with colour; we should see, that every time a man breathed, the air would rush in a stream into his mouth, and then return again; and the air which returned would, being warm, be lighter than the outer air, and would rise upward over the man's head, whe e, cooling and mingling with the outer air, it would descend again. We do, in fact, see this action evidenced: when in winter time the cold condenses the vapour of the breath, we see the little cloud constantly rising before the breather's face, and dispersing in the surrounding air.

Is it not a wonderful thing that that clear and elastic substance, which you cannot feel, though it touches every part of your body, and which you cannot see, is composed of two distinct bodies, having very different properties; and that the two bodies can easily be separated from each other?

Air is of the first importance to life. Hence it is provided for us everywhere. We require air every second, water every few hours, and food at intervals considerably apart. Air is therefore provided for us everywhere. Whether we stand or sit; whether we dwell in a valley or upon a mountain; whether we go into the cellar under our house, or into the garret at the top of it air is there provided for us. God, who made it a law that man should breathe to live, also sent him air bundantly, that he might comply with that law. And all that is required from man in this respect is, that he will not shut out God's bounty, but receive it freely.

As we have employed the idea that if the air were coloured we should have the opportunity of marking the process of breathing, let us enlarge upon this, and suppose that every time the air were returned from the lungs it became of a darker colour, the darkness denoting increasing impurity. If we placed a man in a room full of pure air, we should see the air enter his lungs, and sent back slightly tinged; but this would disperse itself with the other air of the room and scarcely be perceptible. As the man continued to breathe, however, each measure of air returning from the lungs would serve to pollute that abiding in the room, until at last the whole mass would become cloudy and discoloured, and we should see such a change as occurs when water is turned from a pure and clear state into a muddy condition. The air does become polluted with each respiration, and although it is colourless, it is as impure as if with every breath given off from the lungs it became of a dark colour in proportion to its impurity.

Thus we see how important it is that we should provide ourselves with pure air; and that, in seeking warmth and comfort in our houses, we should provide adequate supply of fresh atmosphere—because it is more vital to life than either water or food.

Indeed, so constant is our requirement of air that ***if we had to fetch it, for***

"There is a natural body, and there is a spiritual body."—I CORINTHIANS XV.

purposes of breathing, or simply to raise it to our mouths, as we do water when we drink, it would be the sole occupation of our lives—we could do nothing else. For this reason, God has sent the air to us, and not required us to go to the air. And the great error of man is, that in too many instances, he shuts off the supply from himself, and brings on disease and pain by inhaling a poisonous compound, instead of air of a healthful kind, which bears an adaptation to the wants of life.

Whilst the rooms of our house are filled with air, it is otherwise with water, which we require in less degree than air. If we have not the artificial means by which water is brought to our houses, through the pipes of a water company, there is a spring or a pump in the garden: or in the absence of these, a good sound cask, standing at the end of our house, forming a receptacle to the water-pipes that surround it, provides us with a supply of water distilled from the clouds. If we were to drink a good draught of water once a day, that would be sufficient for all the purposes of life, as far as regards the alimentary uses of water. Man is, therefore, allowed to go to the stream for his drink, and is required to raise it to his lips at those moments when he uses it.

Although, in breathing, man separates the *oxygen* of the air from the *nitrogen* thereof, he does not separate the *oxygen* of the water from the *hydrogen* Water, in fact, undergoes no change in the body, excepting that of admixture with the substances of the body. And its uses are, to moisten, to cool, to cleanse, and also to nourish the parts with which it comes in contact. But it affords no nourishment of itself; it mixes with the blood, of which it forms a material part, and is the means of conveying *the nourishment of the blood* to every part of the system. After it has filled this office, and taken up impurities that required to be removed, it is cast out of the system again, without undergoing any chemical change.

Man's body is to his Soul, in many respects, what a house is to its occupant. But how superior is the dwelling which God erected, to that which man has built. Reader, come out of yourself, and in imagination realise the abstraction of the Soul from the body. Make an effort of thought, and do not relinquish that effort, until you fancy that you see your image seated on a chair before you. And now proceed to ask yourself certain questions respecting your bodily tenement—questions which, perchance, have never occurred to you before; but which will impress themselves the more forcibly upon you, in proportion as you realise for a moment the idea of your Soul examining the body which it inhabits. There sits before you a form of exquisite proportions, with reference to the mode of life it has to pursue—the wants of the Soul for which it has to care, and which it has to guard, under the direction of that Soul, its owner and master.

Over the brows that mark the intellectual front of that fine form, there fall the auburn locks of youth, or the grey hair of venerable age. Each of those hairs is curiously organised. If you take a branch of a tree, and cut it across, you will find curious markings caused by vessels of various structure, all necessary to the existence of the plant. In the centre will be found either a hollow tube, or a space occupied by a soft substance called pith. Each hair of your head is as curiously formed as the branch of a tree, and in a manner not dissimilar, though its parts are so minute that the unaided eye cannot discern them. Every hair has a root, just as a tree has, and through this root it receives its nourishment. As the vessels

"The very hairs of your head are all numbered."—MATTHEW XI.

which feed a plant are always proportionate to the size of the plant itself, how fine must be those vessels which form the roots of the hair, being in proportion to the size of the hair, which is in itself so small that the eye cannot see its structure? The hair is, in fact, an animal plant, growing upon the body in much the same manner that plants grow upon the surface of the earth. But how does this hair grow? Not alone by the addition of matter at its roots, pushing up and enlongating its stem: nourishment passes up through its whole length, and is deposited upon its end, just as the nourishment of a tree is deposited upon its extreme branches. If, after having your hair cut, you were to examine its ends by the microscope, you would discover the abrupt termination left by the scissors. But allow the hair to grow, and then examine it, and you will discover that it grows from its point which in comparison with its former state, is perfect and fine. The reason why the beard is so hard is, that the ends of the hair are continually being shaved off. The hair of the beard, if allowed to grow, would become almost as soft as the hair of the head.

But why is man's head thus covered with hair? For precisely the same reason that a house is thatched—to keep the inmates warm. We might add, also, to give beauty to the edifice. But as beauty is a conventional quality—and if men were without it they would consider themselves quite as handsome as they do now—we will not enlarge upon the argument. Our bald-headed friends, too, might have reason to complain of such a partial hypothesis. The brain is the great organ upon which the health, the welfare, and the happiness of the system depends. The skull, therefore, may be regarded as analagous to the "strong box," the iron chest in which thé merchant keeps his treasure. There is no point at which the brain can be touched to its injury, without first doing violence to the skull. Even the spinal cord runs down the back through a tunnel or tube, formed in a number of strong bones, so closely and firmly jointed together, that they are commonly termed "the back-bone."

Look at the eyebrows. What purpose do they fulfil? Precisely that of a shed, or arch placed over a window to shelter it from rain. But for the eyebrows the perspiration would frequently run from the brow into the eyes, and obscure the sight; a man walking in a shower of rain would scarcely be able to see; and a mariner in a storm would find a double difficulty in braving the tempest.

Now we come to the eye, which is the window of the Soul's abode. And what a window! how curiously constructed! how wisely guarded! In the eyelashes, as well as the eyebrows, we see the hair fulfilling a useful purpose, differing from any already described. The eyelashes serve to keep cold winds, dust, and too bright sun, from injuring or entering the windows of the body. When we walk against the east wind, we bring the tips of our eyelashes together, and in that way exclude the cold air from the surface of the eye; and in the same manner we exclude the dust and modify the light. The eyelashes, therefore, are like so many sentries, constantly moving to and fro, protecting a most important organ from injury. The eyelids are the shutters by which the windows are opened and closed. But they also cleanse the eye, keeping it bright and moist. There are, moreover, in the lids of each eye or window, little glands, or springs, by which a clear fluid is formed and supplied for cleansing the eye. The eye is placed in a socket of the skull, in which it has free motion, turning right or left up or down, to serve the purpose of the

"Thou art of purer eyes than to behold evil, and canst not look on iniquity."
HABAKKUK I.

inhabitant of the dwelling. Of the structure of the eye itself we will not say much, for the engravings will afford a clearer understanding than a lengthy written description. But we would have you examine the formation of the iris of the living eye, the ring which surrounds the pupil. Hold a light to it, and you will find that the iris will contract and diminish the pupil; withdraw the light, and the iris will relax, and the pupil expand, thus regulating the amount of light. The images of external objects are formed upon the retina of the eye, a thin membrane, spread out upon the extremity of a large nerve, which proceeds immediately to the brain, and forms the telegraphic cord by which information is given to the mind, of everything visible going on within the range of sight.

Now, think for a few moments upon the wonderful structure of those windows of the body. Can you fancy, in the walls of your house, a window which protects itself, cleanses itself, and turns in any direction at the mere will of the tenant; and when that tenant is oppressed by excess of light, draws its own curtain, and gives him ease; and when he falls asleep, closes its own shutters, and protects itself from the cold and dust of night, and the instant he awakes in the morning, opens, cleanses itself with a fluid which it has prepared during the night, and kept in readiness; and repeats this routine of duty day after day for half a century, without becoming impaired? Such, nevertheless, is the wonderful structure of the window of the body—the eye.

In some scientific works that have recently been published, curious investigations have been made known. It has been shown that the eye is impressed momentarily, as a photographic plate is impressed by the rays of the sun. But the photography of the eye has this extraordinary quality—that one image passes away, and another takes its place immediately, without confusion or indistinctness. But the most wonderful assertion of all is, that under the excitement of memory these photographic images are restored; and that when, "in our mind's eye," we see the image of some dear departed friend, the retina really revives an image which once fell upon its sensitive surface, and which image has been stored up for many years in the sacred portfolio of its affections!

Another extraordinary assertion is one which comes supported by a degree of authencity that entitles it to consideration. It is said that the eye of a dead man retains an impression of the last picture that fell upon the faithful retina. Dr. Sandford, of America, examined the eye of a man named Beardley, who had been murdered at Auburn and he published in the *Boston Atlas* the following statement:—"At first we suggested the saturation of the eye in a weak solution of atrophine, which evidently produced an enlarged state of the pupil. On observing this, we touched the end of the optic nerve with the extract, when the eye instantly became protuberant. We now applied a powerful lens, and discovered in the pupil, the rude, worn-away figure of a man, with a light coat, beside whom was a round stone, standing or suspended in the air, with a small handle, stuck in the earth. The remainder was debris, evidently lost from the destruction of the optic, and its separation from the mother brain. Had we performed the operation when the eye was entire in the socket, with all its powerful connection with the brain, there is not the least doubt but that we should have detected the last idea and impression made on the mind and eye of the unfortunate man. The picture would evidently be entire; and perhaps we should have had the contour, or better still, the exact figure of the murderer.

"Be not rash with thy mouth, and let not thine heart be hasty to utter anything before God: for God is in heaven, and thou upon earth; therefore let thy words be few."—ECCLESIASTES V.

The last impression on the brain before death is always more terrible from fear than any other cause, and figures impressed on the pupil more distinct, which we attribute to the largeness of the optic nerve, and its free communication with the brain." Whether the supposition, which seems to be supported by the experiment above detailed, be correct or not, it is in no sense more wonderful than the facts which are already known respecting this curious and perfect organ.

The nose is given us for two purposes—to enable us to respire and to smell. As odours arise from the surface of the earth, the cup or funnel of the nose is turned down to meet them. In the nostrils hair again serves a useful purpose. It not only warms the air which enters the nostrils, but it springs out from all sides, and forms an intersecting net, closing the nostrils against dust, and the intrusion of small insects. If by any means, as when taking a sharp sniff, foreign matters enter the nostrils, the nose is armed with a set of nerves which communicate the fact to certain muscles, and the organs of respiration unite with those muscles to expel the intruding substances. In this action, the diaphragm, or the muscle which divides the abdomen from the chest, is pressed down, the lungs are filled with air, the passage by which that air would otherwise escape through the mouth, is closed up, and then, all at once, with considerable force, the air is pressed through the nostrils, to free them from the annoying substance. So great is the force with which this action takes place, that the passage into the mouth is generally pushed open occasioning the person in whom the action takes place, to cry "'tsha!" and thus is formed what is termed a sneeze. As with the eye, so with the nose—innumerable nerves are distributed over the lining membrane, and these nerves are connected with larger nerves passing to the brain, through which everything relating to the sense of smell is communicated.

The nose acts like a custom-house officer to the system. It is highly sensitive to the odour of most poisonous substances. It readily detects hemlock, henbane, monk's hood, and the plants containing prussic acid. It recognises the fœted smell of drains, and warns us not to breathe the polluted air. The nose is so sensitive, that air containing a 200,000th part of bromine vapour will instantly be detected by it. It will recognise the 1,300,000th part of a grain of otto of roses, or the 13,000,000th part of a grain of musk! It tells us in the mornings that our bed-rooms are impure; it catches the first fragrance of the morning air, and conveys to us the invitation of the flowers to go forth into the fields, and inhale their sweet breath. To be "led by the nose," has hitherto been used as a phrase of reproach. But to have a good nose, and to follow its guidance, is one of the safest and shortest ways to the enjoyment of health.

The mouth answers the fourfold purpose of the organ of taste, of sound, of mastication, and of breathing. In all of these operations, except in breathing, the various parts of the mouth are engaged. In eating we use the lips, the tongue, and the teeth. The teeth serve the purpose of grinding the food, the tongue turns it during the process of grinding, and delivers it up to the throat for the purposes of the stomach, when sufficiently masticated. The lips serve to confine the food in the mouth, and assist in swallowing it, and there are glands underneath the tongue, and in the sides of the mouth, which pour in a fluid to moisten the food. And so watchful are those glands of their duty, that the mere imagination frequently causes them to act. Their fluid is required to modify the intensity of different flavours and condiments in which man, with his love of eating, will indulge. Thus, when we eat anything very acid, as a lemon, or anything very irritating, as Cayenne pepper, the effect thereof upon the sensitive nerves of the tongue is greatly modified by a free flow of saliva into the mouth. And if we merely fancy the taste of any such things, those glands are so watchful, that they will immediately pour out their fluid to mitigate the supposed effect.

In speaking, we use the lips, the teeth, the tongue; and the chest supplies air, which, being controlled in its emission, by a delicate apparatus at the mouth of the wind-pipe, causes the various sounds which we have arranged into speech, and by which, under certain laws, we are enabled to understand each other's wants, participate in each other's emotions, express our loves, our hopes, our fears, and glean those facts, the accumulation of which constitutes knowledge, enhances the happiness of man, and elevates him, in its ultimate results above the lower creatures to which the blessing of speech is denied.

The curious structure of the tongue, and the organs of speech, would fill a very interesting volume. The tongue is unfortunately much abused, not only by those who utter foul words, and convert the blessing of speech, which should improve and refine, into a source of wicked and profane language; but it constantly remonstrates against the abuse of food, and the use of things which are not only unnecessary for the good of our bodies, but prejudicial to their health. When the body is sufficiently fed, the tongue ceases its relish, and derives no more satisfaction from eating: but man contrives a variety of inventions to whip the tongue up to an unnatural performance of its duty, and thus we not only over-eat, but eat things that have no more business in our stomachs, than have the stones that we walk upon. Can we wonder, then, that disease is so prevalent, and that death calls for many of us so soon.

That wonderful essence, the Soul of man, rises above all finite knowledge. Its wonders and powers will never, probably, be understood until when, in a future state of existence, the grandest of all mysteries shall be explained. When we talk of the brain, we speak of that which it is easy to comprehend as the organ, or the seat of the mind; when we speak of the mind, we have greater difficulty in comprehending the meaning of the term we employ; but when we speak of the Soul, we have reached a point which defies our understanding, because our knowledge is limited. The brain may be injured by a blow; the mind may be pained by a disagreeable sight, or offended by a harsh word; but the Soul can only be influenced secondarily through the mind, which is primarily affected by the organs of the material senses. Thus the happiness or the misery of the Soul depends to a very great extent upon the proper fulfilment of the duties of the senses, which are the servants of the Soul, over which the mind presides, as the steward who mediates between the employer and the employed

The Ear, which is taught to delight in sweet sounds, and in pure language, is a better servant of the master Soul, than one which delights not in music, and which listens, with approbation or indifference, to the oaths of the profane. The Eye which rejoices in the beauties of nature, and in scenes of domestic happiness and love, is a more faithful servant than one that delights in witnessing scenes of revelry, dissipation, and strife. The Nose which esteems the sweet odour of flowers, or the life-giving freshness of the pure air, is more dutiful to his master than one that rejects not the polluted atmosphere of

"Out of the same mouth proceedeth blessing and cursing. My brethren, these things ought not so to be."—JAMES III.

neglected dwellings. The Mouth which thirsts for morbid gratification of taste, is more worthless than one which is contented with wholesome viands, and ruled by the proper instincts of its duty. Who that can understand the wonderful structure of the tongue, and the complicated mechanism of the organs of speech and of hearing, could be found to take pleasure in the utterance of oaths, and of words of vulgar meaning? Were those beautiful cords that like threads of silk are woven into the muscular texture of the mouth, and along which the essence of life travels with the quickness of thought, to do the bidding of the will—were they given for no higher use than to sin against the God who gave them, and upon whose mercy their existence every moment depends?

The actions of the senses must necessarily affect the mind, which is the head steward of the Soul; and the Soul becomes rich in goodness, or poor in sin, in proportion as the stewardship, held by his many servants, is rightly or wrongfully fulfilled. As in an establishment where the servants are not properly directed and ruled, they often gain the ascendancy, and the master has no power over them, so with man, when he gives himself up to sensual indulgences. The Soul becomes the slave of the senses—the master is controlled by the servants.

With regard to the mechanism of motion, let us take the case of a man who is walking a crowded thoroughfare, and we shall see how active are all the servants of the Soul, under the influence of the mind. He walks along in a given direction. But for the act of volition in the mind, not a muscle would stir. The eye is watching his footsteps. There is a stone in his path, the eye informs the mind, the mind communicates with the brain, and the nerves stimulate the muscles of the leg to lift the foot a little higher, or turn it on one side, and the stone is avoided. The eye alights on a familiar face, and the mind remembers that the eye has seen that face before. The man goes on thinking of the circumstance under which he saw that person, and partially forgets his walk, and the direction of his steps. But the nerves of volition and motion unite to keep the muscles up to their work, and he walks on without having occasion to think continually, "I must continue walking." He has not to make an effort to lift his leg along between each interval of meditation: he walks and meditates the while. Presently a danger approaches him from behind. The eye sees it not—knows no more, in fact, than if it were dead. But the ear sounds the alarm, tells the man, by the rumbling of a wheel, and the tramp of horses' feet, that he is in danger; and then the nerves, putting forth their utmost strength, whip the muscles up to the quick performance of their duty; the man steps out of the way of danger, and is saved. He draws near to a sewer, which is vomiting forth its poisonous exhalations. The eye is again unconscious—it cannot see the poison lurking in the air. The ear, too, is helpless; it cannot bear witness to the presence of that enemy to life. But the nose detects the noxious agent, and then the eye points out the direction of the sewer, and guides his footsteps to a path where he may escape the injurious consequences. A clock strikes, the ear informs him that it is the hour of an appointment; the nerves stimulate the muscles again, and he is hastened onward. He does not know the residence of his friend, but his tongue asks for him, and his ear makes known the reply. He reaches the spot—sits—rests. The action of the muscles is stayed; the nerves are for a time time at rest. The blood which had flown freely to feed the muscles while they were working goes more steadily through the arteries and veins, and the lungs, which had been purifying the blood in its course, partake of the temporary rest.

Let us remember that there are two sets of muscles, acting in unison with each other, to produce the various motions; they are known by the general terms of *flexors* and *extensors;* the first enable us to bend the limbs, the other to bring the limbs back to their former position. The flexors enable us to close the hand, the extensors to open it again. The flexors enable us to raise the foot from the ground· the extensors set the foot down again in the place desired. Consider for a moment the nicety with which the powers of these muscles must be balanced, and the harmony which must subsist between them in their various operations. When we are closing the hand, if the extensor muscles did not gradually yield to the flexors—if they gave up their hold all at once, the hand, instead of closing with gentleness and ease, would be jerked together in a sudden and most uncomfortable manner. If, in such a case, you were to lay your hand with its back upon the table, and wish to close the hand, the fingers would fall down upon the palm suddenly, like the lid of a box. Again, consider how awkward it would be in such a case; our walk through the streets would become a series of jumps and jerks; when a man had raised his foot, after it had been jerked up, there it would stand fixed for a second before the opposite muscles could put on their power to draw it down again. This case is not at all suppositious: there is a derangement frequently observed in horses, in which one set of muscles becomes injured, and we may see horses suffering from this ailment, trotting along with one of their legs jerking up much higher than the others, and set down again with difficulty, just in the manner described.

It is also to be observed that very nice proportions must exist between the sizes of the muscles and the sizes of the bones. If this were not the case, our motions, instead of being firm and steady, would be all shaky and uncertain. In old persons the muscles become weak and relaxed; hence there is a tendency in the movements of the aged to fall, as it were, together; the head is no longer erect, the body bends, the knees totter, and the arms lean towards the body as for support.

In the child a somewhat similar state of things exists. The muscles have not been properly developed, nor have they been brought sufficiently under the controul of the nervous system. The child, therefore, totters and tumbles about, and it is not until it has stumbled and tumbled some hundreds of times in its little history, that the muscles have become strong enough to fulfil their office, or have been brought sufficiently under the controul of the nervous system, to perform well the various duties required from them.

In all these things, we recognise the perfection of the divine works. We are apt, too apt, to overlook this perfection, because it prevails in everything; but by speculating upon what inconveniences we might suffer, were not things ordained as they are, we obtain most convincing evidences of divine goodness and wisdom.

Having taken this view of the muscular system of the external man, let us turn our attention to the muscles of the internal organs. The muscles of which we have been speaking are called the voluntary muscles, because we have them under our own controul—they are subject to the influences of our will. But there is the other set of muscles. What are they? We talk of the beating, or of the palpitation, of the heart. But, what is it that causes the heart to beat? You cannot, if you wish it, make your heart beat more quickly or more slowly. Place your finger upon your pulse, and notice the degree of rapidity with which its pulsations follow. Now think that you should like to double the frequency of those pulsations. Say to the heart, with your inner voice, that you wish it to beat 120 times in a minute, instead of 60. It does not obey you; it does not appreciate your command. Now place your finger on the table, and your watch by the side of your hand, and tell your finger to beat 60 times in the minute, or 100 times, or 150 times, or 200 times, and the finger will obey you—because it is *moved by muscles which are subject to the will*, while the heart is composed of muscles which are *not subject to the will*. Why should this be? Why should man have the power to regulate his finger, and not to regulate his heart?

For the sustentation of our bodies it is needful that the blood should ever be in circulation. If the heart were to cease beating only for three or four minutes (perhaps less) life would be extinct. In this short time the whole framework of man, beautiful in its proportions, perfect in its parts, would pass into the state of dead matter, and would simply wait the decay that follows death. The eye would become dull and glazed, the lips would turn blue, the skin would acquire the coldness of clay—love, hope, joy, would all cease. The sweetest, the fondest ties would be broken. Flowers might bloom, and yield their fragrance, but they would be neither seen nor smelt; the sun might rise in its brightest splendour, yet the eye would not be sensitive to its rays; the rosy-cheeked child might climb the paternal knee; but there, stiff, cold, without joy, or pain, or emotion of any kind, unconscious as a block of marble, would sit the man *whose heart for a few moments had ceased to beat.*

How wise, then, and how good of God, that he has not placed this vital organ under our own care! How sudden would be our bereavements—how frequent our deaths, how sleepless our nights, and how anxious our days, if we had to keep our own hearts at work, and death the penalty of neglect.

And yet, before we were born, until we reach life's latest moment—through days of toil, and nights of rest—even in the moments of our deepest sin against the God who at the time is sustaining us, our hearts beat on, never stopping, never wearying, never asking rest.

This brings us to another reflection. Our arms get weary, our legs falter from fatigue, the mind itself becomes overtaxed, and all our senses fall to sleep. The eye sees not, the ear is deaf to sound, the sentinels that surround the body, the nerves of touch, are all asleep—you may place your hand upon the brow of the sleeping man, and he feels it not. Yet, unseen, unheard, without perceptible motion, or the slightest jar to mar the rest of the sleeper, the heart beats on, and on, and on. As his sleep deepens, the heart slackens its speed, that his rest may be the more sound. He has slept for eight hours, and the time approaches for his awakening. But is the heart weary—that heart which has toiled through the long and sluggard night? No! The moment the waking sleeper moves his arm, the heart is aware that a motion has been made, that effort and exercise are about to begin. The nerves are all arousing to action: the eyes turn in their sockets, the head moves upon the neck; the sleeper leaves his couch, and the legs are once more called upon to bear the weight of the body. Blood is the food of the eye, the food of the ear, of the foot, the hand and every member of the frame. While they labour they must be fed—that is

"Awake up, my glory; awake, psaltery and harp: I myself will awake early."—
PSALM LVII.

the condition of their life, the source of their strength. The heart, therefore, so far from seeking rest, is all fresh and vigorous for the labours of the day, and proceeds to discharge its duty so willingly, that we do not even know of the movements that are going on within us.

Thus we have seen the difference between the voluntary and the involuntary muscles, and we have perceived the goodness of our Creator in not entrusting to our keeping the controul of an organ so vital to life, as the heart.

But the heart is not the only organ which thus works unseen and unfelt. There are the lungs and the muscles of the chest, the stomach, and other parts occupying the abdomen, together with all those muscular filaments which enter into the structure of the coats and valves of the blood-vessels, and which assist to propel the blood through the system. All these are at work at every moment of man's life; and yet, so perfect is this complicated machinery, that we really do not know, except by theory, what is going on within us.

During the time that the sleeper has been at rest, the stomach has been at work digesting the food which was last eaten. Then the stomach has passed the macerated food into the alimentary canal, the liver has poured out its secretion, and produced certain changes in the condition of the dissolved food: and the lacteals, of which there may be many thousands, perhaps millions, have been busy sucking up those portions of the food which they knew to be useful to the system, whilst they have rejected all those useless and noxious matters upon which the liver, like an officer of health, had set his mark, as unfitting for the public use. This busy life has gone on uninterruptedly; every member of that body, every worker in that wonderful factory, has been unremitting in his duty, and yet the owner, the master, has been asleep, and wakes up finding every bodily want supplied!

Notwithstanding that much has already been said of the wonders that pertain to the eye, it has not yet been considered as the seat of *tears*, those mute but eloquent utterers of the sorrows of the heart. Beautiful Tear! whether lingering upon the brink of the eyelid, or darting down the furrows of the care-worn cheek—thou art sublime in thy simplicity—great, because of thy modesty—strong, from thy very weakness. Offspring of sorrow! who will not own thy claim to sympathy? who can resist thy eloquence? who can deny mercy when thou pleadest?

Every tear represents some in-dwelling sorrow preying upon the mind and destroying its peace. The tear comes forth to declare the inward struggle, and to plead a truce against further strife. How meet that the eye should be the seat of tears—where they cannot occur unobserved, but, blending with the beauty of the eye itself, must command attention and sympathy!

Whenever we behold a tear, let our kindliest sympathies awake—let it have a sacred claim upon all that we can do to succour and comfort under affliction. What rivers of tears have flown, excited by the cruel and perverse ways of man! War has spread its carnage and desolation, and the eyes of widows and orphans have been suffused with tears! Intemperance has blighted the homes of millions, and weeping and wailing have been incessant! A thousand other evils which we may conquer have given birth to tears enough to constitute a flood—a great tide of grief. Suppose we prize this little philosophy, *and each one determine never to excite a tear in another.* Watching the eye as the telegraph

"Who is as the wise man? and who knoweth the interpretation of a thing? a man's wisdom maketh his face to shine, and the boldness of his face shall be changed."—ECCLESIASTES VIII.

of the mind within, let us observe it with anxious regard; and whether we are moved to complaint by the existence of supposed or real wrongs, let the indication of the coming tear be held as a sacred truce to unkindly feeling, and our efforts be devoted to the substitution of smiles for tears!

There is only one other matter to which we think it necessary to allude, before we pass to the concluding section of our work. It has been said (162), that snow which is *white*, keeps the earth warm; that *white* as a colour is *cool*, and that *black* absorbs *heat* (23). These assertions may appear to be contradictory, and, taken in connection with the fact of the blackness of the skin of negroes in hot climates, may at a first glance be considered unsatisfactory. They are, however, perfectly reconcileable, and that too, without the slightest evasion of the real bearing of the asserted facts. White snow is warm *on account of its texture*, which, being woolly, forms a layer of non-conducting substance over the surface of the earth, and *keeps in its warmth*; white clothing, worn as a garment consisting of a thin material, is cool, because *the white colour* turns back the rays of the sun that fall upon it. Swansdown, although white, being a non conductor, would be warm, because, though it would reflect the light and heat, it would confine and accumulate the heat of the body. The black skin of the negro is a *living texture*, and is not subject to the same laws that govern dead matter. The skin of the negro is largely provided with cells which secrete a fatty matter that acts as a non-conductor of the *external* heat, and also a much larger number of perspiratory glands than exist in the skins of Europeans. The perspiration cools the blood, and carries off the *internal* heat, while the oily matter gives a shining surface to the skin, and reflects the heat, to which the fatty matter presents itself as a non-conducter. We see, therefore, that there are two express provisions for the cooling of the negroes' skin, independent of the colour. The skin of the Esquimaux who inhabits a cold country is *white*, though it might be supposed that a black skin would best conduce to the warmth of his body. But the Esquimaux has, underneath his skin, *a thick coating of fat*, by which the *internal heat* of the body is prevented from escaping.

This *resume* of the subjects embodied in the form of question and answer in the previous pages, will serve to impress the more important truths upon the mind of the reader, while it has enabled us to fill up many omissions necessitated by the arbitrary form of catechetical composition.

"Ask now the beasts, and they shall teach thee; and the fowls of the air, and they shall tell thee."—JOB XII.

CHAPTER LII.

1029. *Why are there so many bodily forms in the animal creation?*

Because the various creatures which God has created have different modes of life, and the forms of their bodies will be found to present *a perfect adaptation to the lives allotted to them.*

Because, also, the beauty of creation *depends upon the variety of objects of which it consists.* And the greatness of the Creator's power is shown *by the diversity of ends accomplished by different means.*

1030. *Why are birds covered with feathers?*

Because they require a high degree of *warmth*, on account of the activity of their muscles; but in providing that warmth it was necessary that their coats should be of the *lightest material*, so as not to impair their powers of flight; and feathers combine the *highest warming power, with the least amount of weight.*

1031. *Why have ostriches small wings?*

Because, having long legs, they do not require their wings for flight; they are merely used *to steady their bodies while running.*

1032. *Why are ostrich feathers soft and downy?*

Because, as the feathers are not employed for flight, the *strength of the feather as constructed for flying is unnecessary,* and the feathers therefore consist chiefly of a soft down.

1033. *Why have water-birds feathers of a close and smooth texture?*

Because such feathers keep the body of the bird warm and dry, by repelling the water from their surface. A bird could scarcely move through the water, with the downy feathers of the ostrich, because of *the amount of water the down would absorb.*

1034. *Why is man born without a covering?*

Because *man is the only animal that can clothe itself.* As in the various pursuits of life he wanders to every part of the globe, he can adapt himself *to all climates and to any season.*

1035. *Why do the furs of animals become thicker in the winter than in the summer?*

Because the creator has thus provided for the preservation of the warmth of the animals during the cold months of winter.

1036. *Why does a black down grow under the feathers of birds as winter approaches?*

Because the down is a non-conductor of heat, and black the warmest colour. It is therefore best adapted to *keep in* their bodily warmth during the cold of winter.

1037. *Why has man no external appendage to his mouth?*

Because *his hands* serve all the purposes of gathering food, and *conveying it to the mouth.* Man's mouth is simply an *opening;* in other animals it is a *projection.*

1038. *Why have dogs, and other carnivorous animals, long pointed teeth, projecting above the rest?*

Because as they have not hands to seize and controul their food, the projecting teeth enable them to *snap and hold* the objects which they pursue for food.

1039. *Why is the under jaw of the hog, shorter and smaller than the upper one?*

Because the animal pierces the ground *with its long snout,* and then the small under jaw *works freely in the furrow* that has been opened, in quest of food.

1040. *Why have birds hard beaks?*

Because, having no teeth, the beak enables them to *seize, hold,* and *divide their food.*

1041. *Why are the beaks of birds generally long and sharp?*

Because the greater number of birds live by *picking up small objects*, such as worms, insects, seeds, &c. The sharp beak therefore, serves as a *fine pincers*, enabling them to take hold of their food conveniently.

1042. *Why have snipes and woodcocks long tapering bills?*

Because they live upon worms which they find in the soft mud of streams and marshy places; their long bills, therefore, enable them to *dig down into the mud after their prey.*

1043. *Why have woodcocks, snipes, &c., nerves running down to the extremities of their bills?*

Because, as they dig for their prey in the soft sand and mud, they cannot see the worms upon which they live. Nerves are, therefore, distributed to the very point of their bills (where, in other birds, nerves are entirely absent) *to enable them to prehend their food.*

Fig. 67.—SPOONBILL.

1044. *Why have ducks and geese square-pointed bills?*

Because they not only feed by dabbling in soft and muddy soil but they consume a considerable quantity of green food, and their square bills enable them to *crop off the blades of grass.*

"Let the heaven and earth praise him, the seas, and everything that moveth therein."—PSALM LXIX.

1045. *Why has the spoon-bill a long expanded bill, lined internally with sharp muscular points?*

Because the bird *lives by suction*, dipping its broad bill in search of acquatic worms, mollusks, insects and the roots of weeds. The bill forms *a natural spoon*, and the muscular points enable the bird to *filter the mud*, and to retain the nourishment which it finds.

1046. *Why has the spoon-bill long legs?*

Because it *wades in marshy places* to find its food. Its legs are therefore long, for the purpose of keeping its body out of the water, and above the smaller acquatic plants, while it searches for its prey.

1047. *Why have the parrots, &c., crooked and hard bills?*

Because they live upon nuts, the stones of fruit, and hard seeds. The shape of the bill, therefore, enables them to *hold the nut or seed firmly*, and the sharp point enables them to *split or remove the husks.*

1048. *Why can a parrot move its upper as well as its lower bill?*

Because by that means it is enabled to bring the nut or seed nearer the fulcrum, or joint of the jaw. It, therefore, acquires greater power, just as with a pair of nut-crackers we obtain increased power by *setting the nut near to the joint.*

1049. *Why have animals with long necks large throats?*

Animals that graze, or feed from the ground, generally have a more powerful muscular formation of the throat than those which feed in other positions, because a greater effort is required to *force the food upward, that would be needed to convey it down.*

1050. *Why are the bones of birds hollow?*

Because they are thereby rendered *lighter*, and do not interfere with the flight of the bird *as they would do if they were solid.* Greater strength is also obtained by the *cylindrical form of the bone*, and a larger surface afforded for the *attachment of powerful muscles*

"And my hand hath found, as a nest, the riches of the people; and as one gathereth eggs that are left, have I gathered all the earth; and there was none that moved the wing, or opened the mouth, or peeped."—ISAIAH X.

1051. *Why do all birds lay eggs?*

Because, to bear their young in any other manner, would *encumber the body*, and materially interfere with their powers of flight.

As soon as an egg becomes large and heavy enough to be cumbersome to the bird, it is removed from the body. A shell, impervious to air, protects the germ of life within, until from two to twenty eggs have accumulated, and then, although laid at different intervals, their incubation commences together, and the young birds are hatched at the same time.

CHAPTER LIII.

1052. *Why have birds with long legs short tails?*

Because the tails of birds are used to guide them through the air, by *a kind of steerage.* When birds with long legs take to flight, they throw their legs behind, and they then *serve the same purpose as a tail.*

Fig. 68.—PERCH.

1053. *Why have fishes fins?*

The fins of fishes are to them, *what wings and tails are to birds,* enabling them to rise in the fluid in which they live by the *reaction of the motions of the fins upon its substance.*

"Speak to the earth, and it shall teach thee; and the fishes of the sea shall declare unto thee. Who knoweth not in all these that the hand of the Lord hath wrought this."—JOB XII.

1054. *Why are the fins of fishes proportionately so much smaller than the wings of birds?*

Because there is less difference between the *specific gravity* of the body of a fish, and the water in which it moves, than between the body of a bird, and the air on which it flies. The fish, therefore *does not require such an expanded surface to elevate or guide it.*

1055. *Why have fishes scales?*

Because scales, while they afford protection to the bodies of fish, are conveniently adapted to their motions; and as the scales *present no surface to obstruct their passage through the water*, as hair or feathers would do, they evidently form the best covering for the acquatic animal.

1056. *Why do fishes float in streams (when they are not swimming) with their heads towards the stream?*

Because they *breathe* by the transmission of water over the surface of their gills, the water entering at the mouth, and passing over the gills behind. When, therefore, they lie motionless with their heads to the stream, they are in *that position which naturally assists their breathing process.*

1057. *Why have fishes air-bladders?*

Because, as the density of water varies greatly at different depths, the enlargement or contraction of the bladder regulates the relation of *the specific gravity of the body of the fish to that of the water in which it moves.*

1058. *Why have whales a very large development of oily matter about their heads?*

Because their heads are thereby rendered the lighter part of their bodies, and a very slight exertion on the part of the animal will bring its head to the surface *to breathe air, which it constantly requires.*

1059. *Why have birds that swim upon water web-feet?*

Because the spreading out of the toes of the bird brings the membrane between the toes into the form of a fin, or *water-wing*, by striking which against the water, *the bird propels itself along.*

1060. *Why have birds that swim and dive short legs?*

Because long legs would greatly *impede their motions in the water,* by becoming repeatedly entangled in the weeds, and by striking against the bottom. *Waders,* however, require long legs because they have to move about through the *tall vegetation of marshy borders.*

Fig. 69.—STILT-PLOVER AND DUCK.

1061. *Why have the feet of the heron, cormorant, &c., deep rough notches upon their under surface?*

Because, as those birds live by catching fish, they are enabled by the notches in their feet, to *hold the slippery creatures upon which they feed.*

1062. *Why have otters, seals, &c., web-feet?*

Because, while the feet enable them to *walk upon the land,* they are equally effective in their action upon the water, and hence they are *adapted to the amphibious nature of the animals to which they belong.*

1063. *Why do the external ears of animals of prey, such as cats, tigers, foxes, wolves, hyenas, &c., bend forward?*

Because they collect the sounds that occur *in the direction of the*

"Doth the hawk fly by thy wisdom, and stretch her wings toward the south?
"Doth the eagle mount up at thy command, and make her nest on high?

pursuit, and enable the animal to *track its prey* with greater certainty.

1064. *Why do the ears of animals of flight, such as hares, rabbits, deer, &c., turn backward?*

Because they thereby catch the sounds that give them *warning of the approach of danger.*

1065. *Why has the stomach of the camel a number of distinct bags, like so many separate stomachs?*

Because water is stored up in the separate chambers of the stomach, apart from the solid aliment, so that the animal can *feed*, without consuming all its drink. It is thereby *able to retain water to satisfy its thirst while travelling across hot deserts*, where no water could be obtained.

1066. *Why do woodpeckers "tap" at old trees?*

Because by boring through the decayed wood, with the sharp and hard bills with which they are provided, *they get at the haunts of the insects upon which they feed.*

1067. *Why are woodpeckers' tongues about three times longer than their bills?*

Because, if their bills were long, they would not bore the trees so efficiently; and when the trees are bored, and the insects alarmed, they endeavour to retreat into the hollows of the wood; *but the long thin tongue of the woodpecker fixes them on its sharp horny point*, and draws them into the mouth of the bird.

1068. *Why have the Indian hogs large horns growing from their nostrils and turning back towards their eyes?*

Because the horns *serve as a defence to the eyes* while the animal forces its way through the thick underwood in which it lives.

1069. *Why have calves and lambs, and the young of horned cattle generally, no horns while they are young?*

Because the presence of horns would *interfere with the suckling*

"She dwelleth and abideth on the rock, upon the crag of the rock, and the strong place.
"From thence she seeketh the prey, and her eyes behold afar off. Her young ones also suck up blood: and where the slain are, there is she."—JOB XXXIX.

of the young animal. When, however, it is able to feed itself by browsing, *then the horns begin to grow.*

1070. *Why have infants no teeth?*

Because the presence of teeth would interfere with their suckling, while the teeth would be of no service, until the child *could take food requiring mastication.*

1071. *Why cannot flesh-eating animals live upon vegetables?*

Because the gastric juice of a flesh-eating animal, being adapted to the duty which it has to perform, *will not dissolve vegetable matter.*

1072. *Why have birds gizzards?*

Because, having no teeth, the tough and fibrous gizzards are employed *to grind the food preparatory to digestion.*

1073. *Why are small particles of sand, stone, &c., found in the gizzards of birds?*

Because, by the presence of those rough particles, which become embedded in the substance of the gizzard, the food of the bird is more effectively ground.

When our fowls are abundantly supplied with meat, they soon fill their craw, but it does not immediately pass thence into the gizzard; it always enters in small quantities, in proportion to the progress of trituration, in like manner, as in a mill, a receiver is fixed above the two large stones which serve for grinding the corn, which receiver, although the corn be put into it by bushels, allows the grain to dribble only in small quantities into the central hole in the upper mill-stone.—*Paley.*

CHAPTER LIV.

1074. *Why has the mole hard and flat feet, armed with sharp nails?*

Because the animal is thereby enabled to *burrow in the earth,* in search for worms. Its feet are so many *shovels.*

1075. *Why is the mole's fur exceedingly glossy and smooth?*

Because its smoothness enables it to work under ground *without the soil sticking to its coat*, by which its progress would be impeded. From soils of all kinds, the little worker emerges shining and clean.

What I have always most admired in the mole is its *eyes*. This animal occasionally visiting the surface, and wanting, for its safety and direction, to be informed when it does so, or when it approaches it, a perception of light was necessary. I do not know that the clearness of sight depends at all upon the size of the organ. What is gained by the largeness or prominence of the globe of the eye, is width in the field of vision. Such a capacity would be of no use to an animal which was to seek its food in the dark. The mole did not want to look about it; nor would a large advanced eye have been easily defended from the annoyance to which the life of the animal must constantly expose it. How indeed was the mole, working its way under ground, to guard its eyes at all? In order to meet this difficulty, the eyes are made scarcely larger than the head of a corking-pin; and these minute globules are sunk so deeply in the skull, and lie so sheltered within the velvet of its covering, as that any contraction of what may be called the eye-brows, not only closes up the apertures which lead to the eyes, but presents a cushion, as it were, to any sharp or protruding substance which might push against them. This aperture, even in its ordinary state, is like a pin-hole in a piece of velvet, scarcely pervious to loose particles of earth. —*Paley*.

Fig. 70.—ELEPHANTS DRINKING.

1076. *Why has the elephant a short unbending neck?*

Because the elephant's head is so heavy, that it could not have been supported at the end of a long neck (or lever), without *a provision of immense muscular power*.

"Be not afraid, ye beasts of the field: for the pastures of the wilderness do spring, for the tree beareth her fruit, the fig-tree and the vine do yield their strength."—JOEL II.

1077. *Why has the elephant a trunk?*

The trunk of an elephant ***serves as a substitute for a neck***, enabling the animal to crop the branches of trees, or to raise water from the stream.

1078. *Why do the hind legs of elephants bend forward?*

Because the weight of the animal is so great, that when it lay down it would *rise with great difficulty*, if its legs bent outward, as do the legs of other animals. Being bent *under the body*, they have a greater power of pushing directly upward, when the powerful muscles of the thighs straighten them.

According to Cuvier, the number of muscles, in an elephant's trunk, amounts to *forty thousand*, all of which are under the will, and it is to these that the proboscis of this animal owes its flexibility. It can be protruded or contracted at pleasure, raised up or turned to either side, coiled round on itself or twined around any object. With this instrument the elephant collects the herbage on which he feeds and puts it into his mouth; with this he strips the trees of their branches, or grasps his enemy and dashes him to the ground. But this admirable organ is not only adapted for seizing or holding substances of magnitude; it is also capable of plucking a single leaf, or of picking up a straw from the floor. The orifices of the canals of the extremity are encircled by a projecting margin, produced anteriorly into a finger-like process endowed with a high degree of sensibility and exceedingly flexible. It is at once a finger for grasping and a feeler: the division between the two nasal orifices or their elevated sides serves as a point against which to press; and thus it can pick up or hold a small coin, a bit of biscuit, or any trifling thing with the greatest ease.—*Knight's Animal Kingdom.*

1079. *Why have bats hooked claws in their wings?*

Because bats are almost destitute of legs and feet; at least those organs are included in their wings. If they alight upon the ground, they have great difficulty in again taking to the wing, as they cannot run or spring to bring their wings in action upon the air. At the angle of each wing there is placed, therefore, a bony hook, by which the bat attaches itself to the sides of rocks, caves, and buildings, laying hold of crevices, joinings, chinks, &c.; and when it takes its flight, *it unhooks itself, and its wings are at once free to strike the air.*

1080. *Why does the bat fly by night?*

Because it lives chiefly upon moths, which are *night-flying insects*.

"So are the paths of all that forget God; and the hypocrite's hope shall Whose hope shall be cut off, and whose trust shall be a spider's web."—JOB VIII.

1081. *Why does the bat sleep during the winter?*

Because, as the winter approaches, the moths and flying insects upon which it feeds, disappear. *If, therefore, it did not sleep through the winter it must have starved.*

Fig. 71.—BAT WITH HOOKED WINGS.

1082. *Why has the spider the power of spinning a web?*

Because, as it lives upon flies, but is *deficient of the power of flying in pursuit of them,* it has been endowed with an instinct *to spread a snare to entrap them,* and with the most wonderful machinery to give that instinct effect.

There are few things better suited to remove the disgust into which young people are betrayed on the view of some natural objects, than this of the spider. They will find that the most despised creature may become a subject of admiration, and be selected by the naturalist to exhibit the marvellous works of the creation. The terms given to these insects, lead us to expect interesting particulars concerning them, since they have been divided into vagrants, hunters, swimmers, and water spiders, sedentary, and mason-spiders; thus evincing a variety in their condition, activity, and mode of life; and we cannot be surprised to find them varying in the performance of their vital functions (as, for example, in their mode of breathing), as well as in their extremities and instruments. Of these instruments the most striking is the apparatus for spinning and weaving, by which they not only fabricate webs to entangle their prey, but form cells for their residence and concealment; sometimes living in the ground, sometimes under water, yet breathing the atmosphere. Corresponding with their very singular organisation are their instincts. We are familiar with the watchfulness and voracity of some spiders, when their prey is indicated by the vibration of the cords of their net-work. Others have the eye and disposition of the lynx or tiger, and after couching in concealment, leap upon their victims. Some conceal themselves under a silken hood or tube, six eyes only projecting. Some bore a hole in the earth, and line it as finely as if it were done with the trowel and mortar, and then hang it with delicate curtains. A very extraordinary degree of contrivance is exhibited in the trap door spider. This door, from which it derives its name, has a frame and hinge on the mouth of the cell, and is so provided that the claw of the spider can lay hold of it, and

"The spider taketh hold with her hands, and is in king's palaces."—
PROVERBS XXX.

whether she enters or goes out, says Mr. Kirby, the door shuts of itself. But the water-spider has a domicile more curious still: it is under water with an opening at the lower part for her exit and entrance; and although this cell be under water, it contains air like a diving-bell, so that the spider breathes the atmosphere. The air is renewed in the cell in a manner not easily explained. The spider comes to the surface; a bubble of air is attracted to its body; with this air she descends, and gets under her cell, when the air is disengaged and rises into the cell; and thus, though under water, she lives in the air. There must be some peculiar property of the surface of this creature by which she can move in the water surrounded with an atmosphere, and live under the water breathing the air.

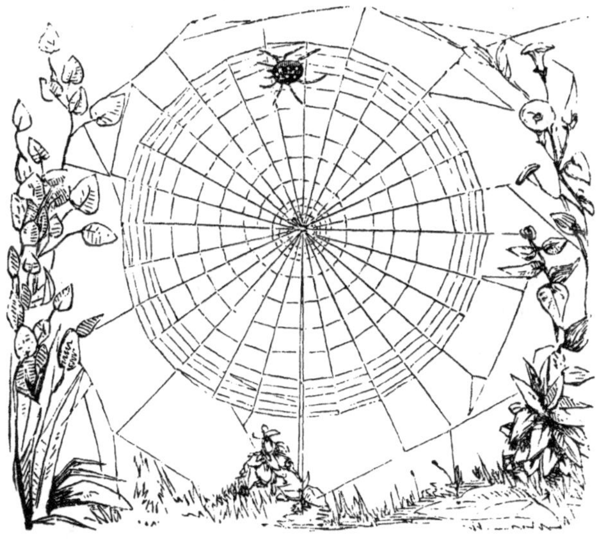

Fig. 72.—WEB OF THE GEOMETRICAL SPIDER.

The chief instrument by which the spider performs these wonders is the spinning apparatus. The matter from which the threads are spun is the liquid contained in cells; the ducts from these cells open upon little projecting teats, and the atmosphere has so immediate an effect upon this liquid, that upon exposure to it the secretion becomes a tough and strong thread. Twenty-four of these fine strands form together a thread of the thickness of that of the silk-worm. We are assured that there are three different sorts of material thus produced, which are indeed required for the various purposes to which the

"For every beast of the forest is mine, and the cattle upon a thousand hills."—
PSALM L.

are applied—as, for example, to mix up with the earth to form the cells; to line these cells as with fine cotton; to make light and floating threads by which they may be conveyed through the air, as well as those meshes which are so geometrically and correctly formed to entrap their prey.—*Note by Lord Brougham to Paley's Natural Theology.*

1083. *Why have many insects a great number of eyes?*

Because the orb of the eye is fixed; there is therefore placed over the eye a multiple-lens, which conducts light to the eye from every direction; so that *the insect can see with a fixed eye as readily as it could have done with a moveable one.* As many as fourteen hundred eyes, or inlets of light, have been counted in the head of a drone-bee. The spider has *eight eyes,* mounted upon different parts of the head; two in front, two in the top of the head, and two on each side.

1084. *Why have birds of prey no gizzards?*

Because their food *does not require to be ground* prior to digestion, as does the food of grain-eating birds.

1085. *Why have earth worms no feet?*

Because the undulatory motion of their muscles serves them for all the purposes of progression *needed by their mode of life.*

1086. *Why have mussels strong tendinous threads proceeding from their shells?*

Because as they live in places that are beaten by the surf of the sea, they *moor their shells* by those threads to rocks and timbers.

1087. *Why have cockles stiff muscular tongues?*

Because, having no threads to moor themselves, as the mussels have, they *dig out with their tongues a shelter for themselves in the sand.*

1088. *Why do oxen, sheep, deer, &c., ruminate?*

Because they have no front teeth in the upper jaw, the place of which is occupied by a hardened gum. The first process, therefore, consists simply of *cropping* their food, which is passed into the paunch, to *be brought up again and ground by the back teeth* when the cropping process is over.

Because, in a wild state they are constantly exposed to the

attacks of carnivorous beasts, and as the mastication of the large amount of vegetable food required for their sustenance would take a considerable time, they are provided with stomachs, by which they are enabled to fill their paunches quickly, and then, retiring to a place of safety, they bring their food up again, and chew it at leisure.

1089. *Why can ruminating animals recover the food from their paunches?*

Because they have a *voluntary power* over the muscles of the throat, by which they can bring up the food at will.

1090. *Why can they keep the unchewed food in the paunch, from the "cud" they have chewed for nourishment?*

Because their stomachs are divided into three chambers: 1, the *paunch*, where the unchewed food is stored; 2, the *reticulum*, where portions of the food are received from the paunch, and moistened and rolled into a "cud," to be sent up and chewed; and 3, the *psalterium*, which receives the masticated food, and continues the process of digestion.

In quadrupeds the deficiency of teeth is usually *compensated* by the faculty of rumination. The sheep, deer, and ox tribe, are without fore-teeth in the upper jaw. These ruminate. The horse and ass are furnished with teeth in the upper jaw, and do not ruminate. In the former class, the grass and hay descend into the stomachs nearly in the state in which they are cropped from the pasture, or gathered from the bundle. In the stomach, they are softened by the gastric juice, which in these animals is unusually copious. Thus softened and rendered tender, they are returned a second time to the action of the mouth, where the grinding teeth complete at their leisure the trituration which is necessary; but which was before left imperfect. I say, the trituration which is necessary; for it appears from experiments, that the gastric fluid of sheep, for example, has no effect in digesting plants, unless they have been previously masticated; that it only produces a slight maceration, nearly as common water would do in a like degree of heat; but that when once vegetables are reduced to pieces by mastication, the fluid then exerts upon them its specific operation. Its first effect is to soften them, and to destroy their natural consistency; it then goes on to dissolve them, not sparing even the toughest parts, such as the nerves of the leaves. I think it very probable, that the gratification also of the animal is renewed and prolonged by this faculty. Sheep, deer, and oxen, appear to be in a state of enjoyment whilst they are chewing the cud. It is then, perhaps that they best relish their food.—Paley.

"I am like a pelican of the wilderness: I am like an owl of the desert. I watch and am as a sparrow alone upon the house top."—PSALM CII.

CHAPTER LV.

1091. *Why do quadrupeds that are vegetable eaters feed so continually?*

Because their food contains but a *small proportion of nutrition*, so that it is necessary to digest a *large quantity* to obtain sufficient nourishment.

1092. *Why do flesh eating animals satisfy themselves with a rapid meal?*

Because the food which they eat is *rich in nutritious matter*, and more readily digestible than vegetable food; it does not, therefore, require the same amount of *grinding with the teeth*.

Fig. 73.—PELICAN WITH DILATED POUCH.

1093. *Why has the pelican a large pouch under its bill*

Because it subsists upon fish, generally of the smaller kind, and uses its pouch *as a net* for catching them; the pouch also serves as

"And God created great whales, and every living creature that moveth, which the waters brought forth abundantly, after their kind, and every winged fowl after his kind: and God saw that it was good."—GENESIS I.

a *paunch*, in which the fish are stored, until the bird ceases from the exertion of fishing, and takes its meal at leisure.

In their wild state they hover and wheel over the surface of the water, watching the shoals of fish beneath, and suddenly sweeping down, bury themselves in the foaming waves; rising immediately from the water by their own buoyancy, up they soar, the pouch laden with the fish scooped up during their momentary submersion. The number of fish the pouch of this species will contain may be easily imagined when we state that it is so dilatable as to be capable of containing two gallons of water; yet the bird has the power of contracting this membranous expansion, by wrinkling it up under the lower mandible, until it is scarcely to be seen. In shallow inlets, which the pelicans often frequent, it nets its prey with great adroitness.

The pelican chooses remote and solitary islands, isolated rocks in the sea, the borders of lakes and rivers, as its breeding place. The nest, placed on the ground, is made of coarse grasses, and the eggs, which are white, are two or three in number. While the female is incubating, the male brings fish to her in his pouch, and the young, when hatched, are assidiously attended by the parents, who feed them by pressing the pouch against the breast, so as to transfer the fish from the former into the throats of the young. This action has doubtless given origin to the old fable of the pelican feeding its young with blood drawn from its own breast.—*Knight's Animal Kingdom.*

1094. *Why do the smaller animals breed more abundantly than the larger ones?*

Because the smaller ones are designed to be the food of the larger ones, and are therefore *created in numbers adapted to that end.* An elephant produces but one calf; the whale but one young one; a butterfly lays six hundred eggs; silk-worms lay from 1,000 to 2,000 eggs; the wasp, 5,000; the ant, 4,000 to 5,000; the queen bee, 5,000 to 6,000, or 40,000 to 50,000 in a season; and a species of white ant *(termes fatalis)* produces 86,400 eggs in a day. Birds of prey seldom produce more than two eggs; the sparrow and duck tribe frequently sit upon a dozen; in rivers there prevail a thousand minnows for one pike; and in the sea, a million of herrings for a single shark; while of the animalcules upon which the whale subsists, there must exist hundreds of millions for one whale.

1095. *Why has the whale feathery-like laminæ of whalebone extending from its jaws?*

Because these feathery bones, lying side by side, form a *sieve, or strainer,* for the large volumes of water which the whale receives into its its mouth, drawing off therefrom millions of small animals which form a jelly-like mass upon which the whale feeds. A whale has been known to weigh as much as 249 tons, and its blubber yielded 4,000 gallons of oil. How many millions of living creatures must have gone to make up that enormous mass of animal matter

1096. *Why have cats, and various other animals, whiskers?*

The whiskers of cats, and of the cat tribe, are exceedingly sensitive, enabling them, when seizing their prey in the dark, to *feel its position most acutely.* These hairs are supplied, through their roots, with branches of the same nerves that give sensibility to the lips, and that in insects *supply their "feelers."*

1097. *Why has the horse a smaller stomach proportionately than other animals?*

Because the horse was created for speed. Had he the ruminating stomach of the ox, he would be quite unfitted for the labour which he now so admirably performs.

1098. *Why has the horse no gall-bladder?*

Because the rapid digestion of the horse, by which its fitness for speed is greatly increased, *does not require the storing up of the bile* as in other animals in which the digestive process is a slower operation.

1099. *Why do certain butterflies lay their eggs upon cabbage leaves?*

Because the cabbage leaves are *the food of the young caterpillars;* and although the butterfly does not subsist herself upon the leaf, she knows by instinct that the leaf will afford food to her future young; she therefore lays her eggs where her young ones will find food.

This explanation applies to many insects that lay their eggs upon other plants

1100. *Why have insects long projections from their heads, like horns or feathers?*

Because those organs (the *antennæ*), are those through which some insects *hear* and others *feel;* and the projecting of these *antennæ* from their bodies probably enables them to hear or feel more acutely while their wings are in motion, *without the interference of the vibrations of their wings.*

1101. *Why have bees stings?*

Because they gather and store up honey which would constantly attract other insects, and the bees would be robbed of their food but for the sting, *which is given to them for protection.*

1102. *Why have flies fine hairs growing at the extremities of their legs?*

Because they require to cleanse their bodies and wings, and to free them from particles of dust. And as they cannot turn their heads for this purpose, they have hairy feet, which serve as brushes, by which any part of their bodies can be reached and cleaned.

CHAPTER LVI.

1103. *Why when the perfume of flowers is unusually perceptible may wet weather be anticipated?*

Because when the air is damp it *conveys the odours of flowers* more effectively than it does when dry.

1104. *Why when swallows fly low may wet weather be expected?*

Because the insects which the swallows pursue in their flight are flying low, *to escape the moisture of the upper regions of the atmosphere.*

1105. *Why do ducks and geese go to the water, and dash it over their backs on the approach of rain?*

Because by wetting the outer coat of their feathers before the rain falls, by sudden dashes of water over the surface, they *prevent the drops of rain from penetrating to their bodies through the open and dry feathers.*

1103. *Why do horses and cattle stretch out their necks and snuff the air on the approach of rain?*

Because they smell the *fragrant perfume* which is diffused in the air by its increasing moistness.

"I will remember the works of the Lord: Surely I will remember thy wonders of old."—PSALM LXXVII.

1107. *Why may change of weather be anticipated when domestic animals are restless?*

Because their skins are exceedingly sensitive to atmospheric influences, and they are oppressed and irritated by *the changing condition of the atmosphere.*

1108. *Why may fine weather be expected when spiders are seen busily constructing their webs?*

Because those insects are highly sensitive to the state of the atmosphere, and when it is setting fine they build their webs, because they know instinctively *that flies will be abroad.*

1109. *Why is wet weather to be expected when spiders hide?*

Because it shows that they are aware that the state of the atmosphere does not *favour the flight of insects.*

1110. *Why if gnats fly in large numbers may fine weather be expected?*

Because it shows that they feel the state of the atmosphere to be favourable, which induces them all to *leave their places of shelter*

1111. *Why if owls scream during foul weather, will it change to fine?*

Because the birds are pleasurably excited by a favourable *change in the atmosphere.*

1112. *Why is it said that the moping of the owl foretels death?*

Because owls scream when the weather is on the change; and when a patient is lingering on a death bed, the alteration in the state of the atmosphere frequently induces death, because the faint and expiring flame of life has not strength enough *to adapt itself to the change.*

1113. *Why may wet weather be expected when spiders break off their webs, and remove them?*

Because the insects, anticipating the approach of rain, remove their webs for preservation.

1114. *Why may we expect a continuance of fine weather when bees wander far from their hives?*

Because the bees feel instinctively that from the state of the atmosphere they may wander far in search of honey, without the danger of being overtaken by rain.

1115. *Why if people feel their corns ache, and their bones rheumatic, may rain be expected?*

Because the dampness of the atmosphere affects its pressure upon the body, and causes a temporary disturbance of the system. All general disturbances of the body, *manifest themselves in those parts which are in a morbid state*—as in a corn, a rheumatic bone, or a decayed tooth.

1116. *Why if various flowers close may rain be expected?*

Because plants are highly sensitive to atmospheric changes, and *close their petals to protect their stamens.*

1117. *Why when moles throw up their hills may rain be expected?*

Because the moles know instinctively, that on the approach of wet, worms move in the ground; the moles therefore become active, *and form their hills.*

1118. *Why is a magpie, when seen alone, said to fortell bad weather?*

Because magpies generally fly in company; but on the approach of wet or cold, one *remains in the nest to take care of the young,* while the other one wanders alone in search of food.

1119. *Why do sea-gulls appear numerous in fine weather?*

Because the fishes swim near to the surface of the sea, and the birds *assemble over the sea to catch the fish, instead of sitting on rocks,* or wading on the shore.

1120. *Why do sea-gulls fly over the land, on the approach of stormy weather?*

Because in stormy weather they cannot catch fish; and the *earth-worms come up on the land* when the rain falls.

"And I said, Oh, that I had wings like a dove! for then would I fly away, and be at rest."—PSALM LV.

1121. *Why if birds cease to sing, may wet, and probably thunder, be expected.*

Because birds are depressed by an unfavourable change in the atmosphere, and *lose those joyful spirits which give rise to their songs.*

1122. *Why if cattle run around in meadows, may thunder be expected?*

Because the electrical state of the atmosphere has the effect of making them feel uneasy and irritable, and *they chase each other about to get rid of the irritability.*

1123. *Why if birds of passage arrive early, may severe weather be expected?*

Because it shows that the indications of unfavourable weather have set in, in the latitudes from which the birds come, and that they have *taken an early flight to escape it.*

1124. *Why if the webs of the gossamer spider fly about in the autumn, may east winds be anticipated?*

Because an east wind is a dry and dense wind, and suitable to the flight of the gossamer spider; the spider feeling instinctively the dryness of the air, throws out its web, and finds it *more than usually buoyant upon the dense air.*

The observation of the changing phenomena which attend the various states of the weather is a very interesting study, though no general rules can be laid down that can be relied upon, because there are modifying circumstances which influence the weather in various localities and climates. To observe weather indications accurately, no phenomenon should be taken alone, but several should be regarded together. The character and the duration of the weather of the preceding days, the direction of the wind, the forms of the clouds, the indications of the barometer, the rise or fall of the thermometer, and the instinctive forewarnings of birds, beasts, insects, and flowers, should all be taken into account. Although no direct material advantages attend such a study, it induces a habit of observation, and developes the inductive faculty of the mind, which, when applied to more significant things, may trace important effects to their greater causes.

Go to the ant, thou sluggard; consider her ways, and be wise.'—PROV. VI.

CHAPTER LVII.

1125. *Why can gossamer spiders float through the air?*

Because, having no wings, and being deficient in the active muscular powers of other spiders, they have been endowed with the power of spinning a web which is so light that it floats in the air, and bears the body of the gossamer spider from place to place. Each web acts as a balloon, and the spider attached thereto is a little *aeronaut.*

1126. *Why do crickets make a peculiar chirping sound?*

Because they have hard wing cases, by the friction of the edges of which they cause their peculiar noise, to *make known to each other where they are,* in the dark crevices in which they hide.

Fig. 74.—GLOW-WORM USING HIS BRUSH.

1127. *Why has the glow-worm a brush attached to its tail?*

Because it is necessary to keep its back very clean, that *the light which its body emits may not be dimmed.*

1128. *Why does the glow-worm emit a light?*

Because the female glow-worm is without wings, but the male is a winged insect. The female, therefore, is endowed with the power of displaying a phosphorescent light. The light is only visible by night, but it is, nevertheless, beautifully adapted for the purpose stated, because *the male is a night flying insect,* and never ventures abroad by day.

There exists some difference of opinion between naturalists upon the uses of the light of a glow-worm; there are some who doubt that it is exhibited to attract the flying insect. The objectors, however, offer no explanation of the luminous properties of the worm. Sir Charles Bell says the preponderance of the argument is decidedly in favour of the explanation we have given.

1129. *Why does not the iris of the fish's eye contract?*

Because the diminished light in water is *never too strong for the retina.*

1130. *Why is the eye of the eel covered with a transparent horny covering?*

Because, as the eel lives in holes, and pushes its head into mud, and under stones, &c., it needed such a covering to *defend the eye.*

1131. *Why is the whale provided with an eye, having remarkably thick and strong coats?*

Because, when he is attacked by the sword-fish and the shark, he is almost helpless against his enemies, as they fix themselves upon his huge carcase. He therefore dives with them down to a depth where the pressure of the water is so great that they cannot bear it. The eye of the whale is expressly organised *to bear the immense pressure of extreme ocean depths,* without impairing the sight.

1132. *Why have fishes no eye-lids?*

Because the water in which they swim keeps their eyes moist. Eyelids would therefore be *useless to them.*

1133. *Why have fishes the power of giving their eye-balls very sudden motion?*

Because, having no eyelids (such organs being unnecessary to keep their eyes moist), they still need the power of freeing their eyes from the contact of foreign matters; and this is secured to them by the power they have of giving the eye-ball a very rapid motion, which causes reaction in the fluid surrounding it, and *sweeps the surface.*

This motion may frequently be seen in the eyes of fishes, in glass globes.

"And God made the beast of the earth after his kind, and cattle after their kind, and everything that creepeth upon the earth after his kind: and God saw that it was good."—GENESIS I.

1134. *Why is the lachrymal secretion of the horse's eye thick and glutinous?*

Because, as his eye is large, and constantly exposed to dust on journies, it is provided with a *viscid secretion*, which cleanses the eye, and more instantly and securely removes the dust, than a *watery* secretion would.

1135. *Why does the lower bill of the sea-crow project beyond the upper one?*

Because the bird obtains his food by *skimming along the water*, into which he dips his bill, and lifts his food out.

1136. *Why do the mandibles of the cross-bill overlap each other?*

Because the bird requires a peculiar bill, to enable it to *split seeds into halves*, and to tear the open cones of the fir-tree.

1137. *Why are the tails of fishes so much larger than their fins?*

Because their tails are their *chief instruments of motion*, while their fins are employed simply to direct their progress, and steady their movements.

1138. *Why have oxen, and other quadrupeds a tough ligament called the "pax-wax," running from their backs to their heads?*

Because their heads are of considerable weight; and having frequent occasion to lift them, they are provided with an elastic ligament, which is fastened at the middle of their backs, while its other extremity is attached to the head. This enables them to raise their heads easily; otherwise the effort to do so would be a work of great labour. To the horse, the pax-wax acts as a natural bearing-rein, assisting it to hold its head in that position which adds to the grace and beauty of the animal.

In carving beef, this ligament may be seen passing along the vertebræ of the neck, the chuck, and the fore ribs.

"He shall feed his flock like a shepherd; he shall gather the lambs with his arm, and carry them in his bosom, and shall gently lead those that are with young."—ISAIAH XL.

1139. *Why have the females of the kangaroo and opossum tribes pouches, or pockets, formed in the skin of their breasts for the reception of their young?*

Because their young ones are remarkably *small and helpless*; in fact, more so than those of any other animal of equal proportions. Besides which, the full grown animals have very long hind-legs, and they progress by a series of extraordinary leaps. It would consequently be impossible for their helpless young ones to follow them: God has therefore given to female kangaroos and opossums curious pockets, *formed out of their own skin*, in which they place their little young ones, and bear them through their surprising leaps with the greatest ease and safety.

CHAPTER LVIII.

1140. *What is the difference between an animal, a plant, and a mineral?*

The great naturalist, Linnæus, used to say that minerals, and animals *grow*, *live*, and *feel*; plants *grow* and *live*; and minerals *grow*.

Animals are here defined to enjoy *three* conditions of existence; plants *two* conditions; and minerals *one* condition.

This definition has, in latter days, been held to be unsatisfactory, since there *are a few plants* that are *supposed to feel*, and *a few animals* that are supposed to have even *less feeling* than the *sensitive plants* alluded to.

The concise definition by Linnæus, nevertheless, is true, as far as regards a *vast majority* of the bodies constituting the three great kingdoms of nature. And it may be sufficient to say that

Animals—grow, live, feel, and move.

Plants—grow and live.

Minerals—grow, by the addition of particles of inorganic matter.

If we now state the few exceptions that are admitted to this definition we shall bring the explanation as near to the truth, as the present state of knowledge will permit.

"And God said, Behold, I have given you every herb bearing seed, which is upon the face of all the earth, and every tree, in the which is the fruit of a tree, yielding seed; to you it shall be for meat."—GENESIS I.

1141. *Why is it understood that some plants feel?*

Because the ***sensitive plant*** closes its leaves on being touched; the *Venus's fly trap* closes its leaves upon flies that alight upon them; others *close* upon the approach of rain, and at sunset, and *open* at sunrise, and turn towards the sun during its daily transit.

1142. *Why is it understood that some plants move?*

Because certain ***sea-weeds*** throw off undeveloped young plants, which move through the water by the aid of fine ***cilia***, or muscular hairs, until they find a suitable place upon which to attach themselves.

The roots of plants will penetrate through the ground in the direction of water, and of favourable soil.

1143. *Of what elementary substances are plants composed?*

Of carbon, oxygen, hydrogen, and nitrogen.

1144. *Whence do plants derive those substances?*

From the air, the earth, and water.

1145. *How do plants obtain carbon?*

They obtain it chiefly from the air, in the form of ***carbonic acid gas.**** The carbon, of the carbonic acid gas, which is thrown out by the breath of animals, and by other processes in nature, is ***absorbed by the leaves of plants***, and the ***oxygen*** which had united with the carbon to form the ***carbonic acid gas***, is again set free for the use of animals.

1146. *How do plants obtain oxygen?*

They obtain it from the ***atmospheric air.*** But as they do not require a large amount of oxygen for their own use, ***they throw off the amount which is in excess***, after having separated it from the other elements with which it was combined when taken up by them. From the humble blade of grass, to the stately tree of the forest, plants operate to purify the air, and to correct and counteract the corruption of the air, by the myriads of animals inhabiting the earth.

It has been generally stated that plants in rooms purify the air by absorbing carbonic acid *by day*, and releasing a part of the oxygen; but that, as the presence of light is necessary to produce this action, they do not restore oxygen to the air, by night, but, on the contrary, give off carbonic acid gas. Therefore it has been stated that plants in rooms by night are unhealthy. Mr. Robert Hunt, one of the ablest chemists of the present time, makes the following remarks upon this subject in his "Poetry of Science:"—

"The power of decomposing carbonic acid is a vital function which belongs to the leaves and bark. It has been stated, on the authority of Leibig, that during the night the plant acts only as a mere bundle of fibres—that it allows of the circulation of carbonic acid and its evaporation, unchanged. In his eagerness to support his chemical hypothesis of respiration, the able chemist neglected to enquire if this was absolutely correct. The healthy plant never ceases to decompose carbonic acid during one moment of its existence; but during the night, when the excitement of light is removed, and the plant reposes, its vital powers are at their minimum of action, and a much less quantity is decomposed than when a stimulating sun, by the action of its rays, is compelling the exertion of every vital function."

In hot, swampy countries, where vegetation is very rapid, and the soil loaded with decomposing carbonic matter, the plants absorb more carbonic acid than they require, and they *then* evolve carbonic acid gas from their leaves. Hence such climates as the West Indies are injurious to *life*, though favourable to *vegetation*.

1147. *How do plants obtain hydrogen?*

They obtain ***hydrogen*** in combination with ***oxygen*** in water, and with ***nitrogen***, in the form of ***ammonia***, as which it exists in animal manures.

1148. *How do plants obtain nitrogen?*

From the ***atmospheric air***, and from the ***soil***, in which it is combined with other elements.

1149. *How do plants apply these elements to the formation of their own structures?*

When those substances which form the food of plants are absorbed, either by their leaves or their roots, they are converted, with the aid of water, into a ***nutritive sap***, which answers the same purposes in ***plants*** as ***blood*** does in ***animals***.

1150. *How is the nutritive sap applied to the growth and enlargement of the plant?*

Every seed contains a small amount of nutrition, sufficient for the sustentation of the ***germ of the plant***, until those vessels are formed, by which the nutritive elements can be absorbed and used for the further development of the living structure.

The earth, penetrated by the sun's rays, warms the sleeping germ, and quickens it into life. For a short time the germ lives upon the seed, which, moistened and warmed by the soil, yields a kind of glutinous sap, out of which the first members of the plant are formed. And then the tender leaf, looking up to the sky, and the slender rootlet penetrating the soil, begin to draw their sustenance from the vast stores of nature.

1151. *Of what do vegetable structures consist?*

Of *membranes*, or thin tissues, which, being variously arranged, form cells, tubes, air passages, &c. Of *fibres*, which form a stronger kind of membrane, and which is variously applied to the production of the organs of the plants. And of *organs*, formed by those elementary substances, by which the plants absorb, secrete, and grow, and fulfil the conditions of their existence.

1152. *Why are seeds generally enveloped in hard cases?*

Because the covering of the seed, like the shell of an egg, is designed *to preserve the germ* within from the influence of external agencies, until the time for development has arrived, and the conditions of germination are fulfilled.

1153. *Why does a seed throw out a root, before it forms a leaf?*

Because moisture, which the root absorbs from the earth, is necessary to enable the germ *to use the nutrition which the seed itself contains*, and out of which the leaf must be eliminated. Moisture forms a kind of gluten, in which the starch of the seed is dissolved, and converted into sugar, the sugar into carbonaceous sap, and the sap into cellular tissue and woody fibre, as the leaves present themselves to the influence of the air and light.

1154. *Why does a plant grow?*

Because, as soon as membranes and vessels are organised in the young germ, the nutritive fluid, formed by its first organs, *begins to move through the fine structures*, and from that time the plant commences to incorporate with its own substance the elements with which it is surrounded, that are suitable to its development.

"Can the rush grow up without mire? can the flag grow without water? Whilst it is yet in his greenness, and not cut down, it withereth before any other herb."—JOB VIII.

CHAPTER LIX.

1155. *Why, if we break the stem of a hyacinth, do we see a glutinous fluid exude?*

Because, by breaking the stem, we rupture the vessels of the plant, and cause the nutritive fluid to escape. The sap of the plant is *analagous to the blood of man,* and the vessels, to the arteries and veins of the animal body.

1156. *Why, if we split the petal of a tulip, do we see cells containing matter of various colours?*

Because, by splitting the petal of the flower, we disclose the anatomy of its structure, and bring to view those cells, or organs, of the vegetable body, by which *the different colouring matters are secreted.*

1157. *Why, if we break a pea-shell across, do we discover a transparent membrane which may be removed from the green cells underneath?*

Because we separate from the cellular, or fleshy part of the shell, the membrane, *which forms the epidermis,* and answers to the skin of the animal body.

1158. *Why, if we cut through a cabbage stump, do we find an outer coat of woody fibre, and an inner substance of cellular matter?*

Because the woody fibre *forms a kind of skeleton,* which supports the internal stricture of the plant, and gives form and character to its organisation. The woody fibre of plants is analagous to the bony structure of animal bodies.

1159. *Why, if we cut across the stem of a plant do we see umerous tubes arranged in parallel lines?*

Because we thereby bring to view *the vessels formed by the membranes and fibres* of the vegetable body, for the transmission of the fluids, by which the structure is sustained.

Skeleton leaves, and seed vessels of plants, form exceedingly interesting objects, and serve to illustrate the wonderful structure of plants. With patience and care, they may be produced by any person, and will afford an interesting occupation. The leaves should be gathered when they are in perfection—that is, when some of the earliest leaves begin to fall from the trees. Select perfect leaves, taking care that they are not broken, or injured by insects. Lay them in pans of *rain water*, and expose them to the air to undergo decomposition Renew the water from time to time, taking care not to damage the leaves. They need not be examined more than once a week, and then only to see that the water is sufficient to cover them. Give them sufficient time for their soft parts to become decomposed, then take them out, and laying them on a white plate with a little water, wash away carefully, with a camel-hair pencil, the green matter that clings to the fibres. The chief requirement is *patience* on the part of the operator, to allow the leaves and seed vessels sufficient time to decompose. Some leaves will take a few weeks, and others a few months, but a large panful may be put to decompose at the same time, and there will always be some ready for the process of cleansing. When they are thorougly cleaned, they should be bleached, by steeping for a short time in a weak solution of chloride of lime. They should then be dried, and either pressed flat, or arranged in boquets for preservation under glass shades. The result will amply reward the perseverance of the operator.

1160. *Why are clayey soils unfavourable to vegetation?*

Because the soil is *too close and adhesive* to allow of the free passage of air or water to the roots of the plants; it also obstructs the expansion of the fibres of the roots.

1161. *Why are sandy soils unfavourable to vegetation?*

Because they consist of particles that have *too little adhesion to each other*; they do not retain sufficient moisture for the nourishment of the plants; and they allow too much solar heat to pass to the roots.

1162. *Why are chalk soils unfavourable to vegetation?*

Because they do not absorb the solar rays, *and are therefore cold to the roots of plants.*

1163. *Why are mixed soils favourable to vegetation?*

Because they contain the *elements of nutrition* essential to the development of the vegetables, and the plants absorb from them those constituents which are necessary to their growth.

1164. *Why do farmers sow different crops in rotation?*

Because every plant takes something from the soil, and gives

"He watereth the hills from his chambers; the earth is satisfied with the fruit of thy works."—PSALM CIV.

something back; but all kinds of plants do not absorb nor restore the elements in the same proportions. Therefore a succession of crops of one kind would soon impoverish the soil; but a succession of crops of different kinds will compensate the soil, in some degree, for the nourishment withdrawn.

1165. *Why do farmers manure their lands?*

Because, as soils vary, and crops impoverish the soils, the farmer employs manure *to restore fertility*, and to *adapt the soils to the wants of the plants* he desires to cultivate.

It is remarkable that Nature herself points out to man the necessity for changing the succession of vegetable growths.

When plants have exhausted the soil upon which they grow, they will push their roots far in search of sustenance, and in time migrate to a new soil, while other plants will spring up and thrive upon the area vacated. When a forest in North America is destroyed by fire, the trees that grow afterwards are unlike those that the fire consumed, and evidently arise from seeds that have long lain buried in the earth, waiting the time when the ascendancy of the reigning order of plants should cease.

1166. *Why are grasses so widely diffused throughout nature?*

Because they form the *food* of a very large portion of the animal kingdom. They have therefore been abundantly provided. No spot of earth is allowed to remain idle long. When the foot of man ceases to tread down the path, grass immediately begins to appear; and by its universality and the hardihood of its nature, it clothes the earth as with a carpet.

Many grasses, whose leaves are so dry and withered that the plants appear dead, revive and renew their existence in the spring by pushing forth new leaves from the bosom of the former ones.—*Withering's Botany.*

Grasses are Nature's care. With these she clothes the earth; with these she sustains its inhabitants. Cattle feed upon their leaves; birds upon their smaller seeds; men upon the larger; for, few readers need be told that the plants which produce our bread-corn, belong to this class. In those tribes which are more generally considered as grasses, their extraordinary means and powers of preservation and increase, their hardiness their almost unconquerable disposition to spread, their faculties of reviviscence, coincide with the intention of nature concerning them. They thrive under a treatment by which other plants are destroyed. The more their leaves are consumed, the more their roots increase. The more they are trampled upon, the thicker they grow. Many of the seemingly dry and dead leaves of grasses revive, and renew their verdure in the spring. In lofty mountains, where the summer heats are not sufficient to ripen the seeds, grasses abound which are viviparous, and consequently able to propagate themselves without seed. It is an observation, likewise, which has often been made, that herbivorous animals attach themselves to the leaves of grasses; and, if at liberty in their pastures to range and choose, leave untouched the straws which support the flowers.—*Paley.*

"For the earth bringeth forth fruit of herself; first the blade, then the ear, after that the full ear in the corn."—MARK V.

CHAPTER LX.

1167. *Why do some plants droop, and turn to the earth after sunset?*

Because, when the warmth of the sun's rays is withdrawn, they turn downwards, and *receive the warmth of the earth by radiation.*

1167* *Why does the young ear of corn first appear enfolded in two green leaves?*

Because the light and air would *act too powerfully for the young ear;* two leaves therefore join, and embrace the ear, and protect it until it has acquired strength, when they divide, and leave the ear to swell and ripen.

1168. *Why are the seeds of plants usually formed within the corollas of flowers?*

Because the petals of the flowers, surrounding the seeds, *afford them protection until they are ripened,* when the flower dies, and the petals fall to the ground.

1169. *Why does the flower of the poppy turn down during the early formation of seed?*

Because the heat would probably be too great for the seed in its early stage. The plant is therefore provided with a curious *curve in its stalk,* which turns the flower downward. But when the seeds are prepared for ripening, *the stalk erects itself,* and the seeds are then presented to the ripening influences of the sun.

1170. *Why have plants of the pea tribe, a folding blossom called the "boat," or "keel?"*

Because, within that blossom the pea is formed, and the shape of the blossom is exactly suited to that of the pea which is formed therein. The blossom is itself protected by external petals; and when the wind blows, and threatens to destroy the parts upon which the seeds depend, the plants *turn their backs to the wind,* and shelter the seed.

"The fruit of the righteous is a tree of life; and he that winneth souls is wise."—PROVERBS XI.

1171. *Why are the leaf buds enclosed in scales which fall off as the leaf opens?*

Because the scales *serve as a shelter* to the tender structure of the young leaf. The scales are rudimentary leaves, formed at the end of the previous season, and which, being undeveloped then, serve to guard the young leaves of the future year.

In trees, especially those which are natives of colder climates, this point is taken up earlier. Many of these trees (observe in particular the *ash* and the *horse-chestnut*) produce the embryos of the leaves and flowers in one year, and bring them to perfection the following. There is a winter therefore to be gotten over. Now what we are to remark is, how nature has prepared for the trials and severities of that season. These tender embryos are, in the first place wrapped up with a compactness, which no art can imitate; in which state they compose what we call the bud. This is not all. The bud itself is enclosed in scales; which scales are formed from the remains of past leaves, and the rudiments of future ones. Neither is this the whole. In the coldest climates, a third preservative is added, by the bud having a *coat* of gum or resin, which, being congealed, resists the strongest frosts. On the approach of warm weather this gum is softened, and ceases to be an hinderance to the expansion of the leaves and flowers. All this care is part of that system of provisions which has for its object and consummation, the production and perfecting of the seeds.—*Paley.*

1172. *Why are the seeds of many plants enclosed in a rich juice, or pulp?*

Because the matter by which the seed is surrounded, as well as being intended for the *nourishment and care of the seed,* is designed for the use of man and of animals, by whom the seed is set free to take its place in the earth.

By virtue of this process, so necessary, but so diversified, we have the seed, at length, in stone-fruits and nuts, incased in a strong shell, the shell itself enclosed in a pulp or husk, by which the seed within is, or hath been, fed; or, more generally (as in grapes, oranges, and the numerous kinds of berries), plunged overhead in a glutinous syrup, contained within a skin or bladder; at other times (as in apples and pears) embedded in the heart of a firm fleshy substance; or (as in strawberries) pricked into the surface of a soft pulp.

These and many more varieties exist in what we call *fruits.* In pulse, and grain, and grasses; seeds (as in the pea tribe) regularly disposed in parchment pods, which, though soft and membranous, completely exclude the wet even in the heaviest rains; the pod also, not seldom, (as in the bean), lined with a fine down; at other times (as in the senna) distended like a blown bladder; or we have the seed enveloped in wool (as in the cotton-plant), lodged (as in pines) between the hard and compact scales of a cone, or barricadoed (as in the artichoke and thistle) with spikes and prickles; in mushrooms, placed under a pent-house; in ferns, within slits in the back part of the leaf; or (which is the most general organisation of all) we find them covered by strong, close tunicles, and attached to the stem according to an order appropriated to each plant, as is seen in the several kinds of grains and of grasses.

In which enumeration, what we have first to notice is, unity of purpose under variety of expedients. Nothing can be more *single* than the design; more *diversified* than the means. Pellicles, shells, pulps, pods, husks, skin, scales armed with thorns, are all employed in prosecuting the same intention. Secondly; we may observe, that in all these cases, the purpose is fulfilled within a just and *limited* degree. We can perceive, that if the seeds of plants were more strongly guarded than they are, their greater security would interfere with other uses. Many species of animals would suffer, and many perish, if they could not obtain access to them. The plant would overrun the soil; or the seed be wasted for want of room to sow itself. It is, sometimes, as necessary to destroy particular species of plants, as it is, at other times, to encourage their growth. Here, as in many cases, a balance is to be maintained between opposite uses. The provisions for the preservation of seeds appear to be directed, chiefly against the inconstancy of the elements, or the sweeping destruction of inclement seasons. The depredation of animals, and the injuries of accidental violence, are allowed for in the abundance of the increase. The result is, that out of the many thousand different plants which cover the earth, not a single species, perhaps, has been lost since the creation.

When nature has perfected her seeds, her next care is to disperse them. The seed cannot answer its purpose, while it remains confined in the capsule. After the seeds therefore are ripened, the pericarpium opens to let them out, and the opening is not like an accidental bursting, but for the most part, is according to a certain rule in each plant. What I have always thought very extraordinary; nuts and shells, which we can hardly crack with our teeth, divide and make way for the little tender sprout which proceeds from the kernel. Handling the nut, I could hardly conceive how the plantule was ever to get out of it. There are cases, it is said, in which the seed-vessel, by en elastic jerk, at the moment of its explosion, casts the seeds to a distance. We all, however, know, that many seeds (those of most composite flowers, as of the thistle, dandelion, &c.) are endowed with what are not improperly called *wings;* that is, downy appendages, by which they are enabled to float in the air, and are carried oftentimes by the wind to great distances from the plant which produces them. It is the swelling also of this downy tuft within the seed-vessel that seems to overcome the resistance of its coats, and to open a passage for the seed to escape.

But the *constitution* of seeds is still more admirable than either their preservation or their dispersion. In the body of the seed of every species of plant, or nearly of every one, provision is made for two grand purposes: first, for the safety of the *germ;* secondly, for the temporary support of the future plant. The sprout, as folded up in the seed, is delicate and brittle beyond any other substance. It cannot be touched without being broken.

Yet in beans, peas, grass-seeds, grain, fruits, it is so fenced on all sides, so shut up and protected, that whilst the seed itself is rudely handled, tossed into sacks, shovelled into heaps, the sacred particle, the miniature plant remains unhurt. It is wonderful, also, how long many kinds of seeds, by the help of their integuments, and perhaps of their oils, stand out against decay. A grain of mustard-seed has been known to lie in the earth for a hundred

"Say not ye, There are four months, and then cometh harvest? behold, I say unto you, Lift up your eyes, and look on the fields; for they are white already to harvest."—JOHN IV.

years; and as soon as it had acquired a favourable situation, to shoot as vigorously as if just gathered from the plant. Then, as to the second point, the temporary support of the future plant, the matter stands thus. In grain, and pulse, and kernels, and pipins, the germ composes a very small part of the seed. The rest consists of a nutritious substance, from which the sprout draws its aliment for some considerable time after it is put forth; viz., until the fibres, shot out from the other end of the seed, are able to imbibe juices from the earth, in a sufficient quantity for its demand. It is owing to this constitution that we see seeds sprout, and the sprouts make a considerable progress, without any earth at all.

From the conformation of fruits alone, one might be led, even without experience, to suppose, that part of this provision was destined for the utilities of animals. As limited to the plant, the provision itself seems to go beyond its object. The flesh of an apple, the pulp of an orange, the meat of a plum, the fatness of the olive, appear to be *more* than sufficient for the nourishing of the seed or kernel. The event shows, that this redundancy, if it be one, ministers to the support and gratification of animal natures; and when we observe a provision to be more than sufficient for one purpose, yet wanted for another purpose, it is not unfair to conclude that both purposes were contemplated together.—*Paley.*

1173. *Why have climbing plants tough curly tendrils?*

Because, *having no woody stalks of their own* to support them, they require to take hold of surrounding objects, and raise themselves from the ground by climbing. Their spiral tendrils are, therefore, so many hands, assisting them to rise from the earth.

1174. *Why does the pea put forth tendrils, and the bean not?*

Because the bean has in its stalk *sufficient woody fibre to support itself*, but the pea has not. We do not know a single tree or shrub having a firm strong stem sufficient for its support which is *also* supplied with tendrils.

1175. *Why do the ears of wheat stand up by day, and turn down by night?*

Because, when the ear is becoming ripe, the cold dew falling into the ear, might *induce blight;* the ears therefore turn down to the earth, and *receive warmth by radiation.*

1176. *Why have grasses, corn, canes, &c., joints, or knots in their stalks.*

Because a long hollow stem would be liable to bend and break. But the joints are so many points where the fibres are bound together, and the structure *greatly strengthened*

"Then shall the earth yield her increase; and God, even our own God, shall bless us."—PSALM XLVII.

1177. *Why have the berries of the mistletoe a thick viscid juice?*

Because the mistletoe is a *parasitical* plant, growing upon the bark of other trees. It will not grow in the ground; its seeds are therefore filled with an exceedingly sticky substance, which serves to attach them to the bark of trees, to which the berries attach themselves at once, by throwing out tough fibres; and the next year the plant grows.

Fig. 75.—THE MISTLETOE.

1178. *How are the seeds of the mistletoe transferred from its own stem to the bark of trees.*

Various birds, and particularly the *missel thrush*, feed upon the berries. As the bird moves in pursuit of its food, the viscid berries attach themselves to its feathers, and in this way the thrush is the instrument which conveys the seed to the spot to which it adheres, and from which the tree ultimately grows.

1179. *What is the circulation of the sap in plants?*

The circulation of the sap is the movement of the nutritive juices by which the plant is sustained. There is a slow uninterrupted movement of the sap from the root through the stems to the leaves, and downwards from the leaves through the bark to the root.

1180. *Why does the sap of plants thus ascend and descend?*

Because it *conveys upward* from the ground some of the matter by which the plant is to be nourished, and which must undergo digestion in the leaves; and it *brings downward* from the leaves the matters absorbed, for the nourishment of the plant, and discharges through the root the substances which the plant cannot use.

The movement of the sap is most active in the spring; but in the depths of the winter it almost ceases.

There are other motions of the sap in plants, which are called *special*, in distinction from the ascending and descending of the sap, which is called *general*, or common to all plants. The special movements of the sap are peculiar to certain plants, in some of which a fluid, full of little green cells, is found to have a rotatory motion; in other plants, a milky fluid is found to move through particular tissues of the vegetable structure.

1181. *Why are the leaves of plants green?*

Because they secrete a carbonaceous matter, named *chlorophyl*, from which they derive their green colour.

1182. *Why are the hearts of cabbages, lettuces, &c., of a pale yellow colour?*

Because the action of *light* is necessary to the formation of *chlorophyl;* and as the leaves are folded upon each other, they exclude the light, and the green matter is not formed.

1183. *Why do leaves turn brown in the autumn?*

Because, when their power of decomposing the air declines, the *oxygen* absorbed in the carbonic acid gas, *lodges in the leaf*, imparting to it a red or brown colour.

1184. *Why do succulent fruits, such as gooseberries, plums, &c., taste acid?*

Because, in the formation of juices, a considerable amount of *oxygen* is absorbed, and the oxygen imparts acidity to the taste.

1185. *Why do ripe fruits taste sweet, and unripe fruits taste sour?*

Because the juices of the ripe fruit contain a large proportion of *sugar*, which in the unripe fruit has not been formed.

1186. *Why do some leaves turn yellow?*

Because they retain an excess of *nitrogen*. Leaves undergoing decay turn either yellow, red, crimson, or violet. Yellow is due to the excess of *nitrogen;* red and crimson to various proportions of *oxygen;* violet to a mixture of *carbon;* and green to *chlorophyl.*

1187. *Why do leaves fall off in the autumn?*

Because they have supplied for a season the natural wants of the tree. Every part has received nutrition through the spring and summer months; and the wants of the tree being supplied, the chief use of the leaf ceases, and it falls to the ground to decay, and enrich the soil.

1188. *Why do plants suffer from the smoke of cities?*

Because the smoke *injures the porous structure of the leaves*, and interferes with their free respiration

LESSON LXI.

1189. *Why are vegetable productions so widely diffused?*

Because they everywhere form the *food of the animal creation.* Without them, neither man nor beast could exist. Even the flesh-eating animals are sustained by them, since they live by preying upon the bodies of vegetable-eaters.

They also enrich and beautify the earth. They present the most charming diversities of proportions and features. From the cowslip, the primrose, and the blue-bell of our childish days, to the broad oak under which we recline, while children gambol round us, they are all beautiful or sublime, and eminently useful in countless ways to man.

They spread a carpet over the surface of the earth; they cling to old ruins, and cover hard rocks, as though they would hide decay, and give warmth to the coldness of stone. In tropical climates they supply rich fruits full of cool and refreshing juices, and they spread out upon the crests of tall trees those broad leaves which shelter the native from the scorching heat of the sun.

They supply our dwellings with furniture of every kind, from the plain deal table, to the handsome cabinet of satin or rosewood; they afford rich perfumes to the toilette, and luscious fruits and wines to the desert; they charm the eye of the child in the daised field; they adorn the brow of the bride; they are laid in the coffin with the dead; and, as the cypress or the willow bend over our graves, they become the emblems of our grief.

1190. *What is mahogany?*

Mahogany is the wood of trees brought chiefly from South America and Spain. The finest kind is imported from St. Domingo, and an inferior kind from Honduras.

We all know the beauty of mahogany wood. But we do not all know that mahogany was first employed in the repair of some of Sir Walter Raleigh's ships at Trinidad in 1597. The discovery of the beauty of its grain for furniture and cabinet work was accidental. Dr. Gibbons, a physician of eminence, was building a house in King-street, Covent-garden; his brother, captain of a West Indiaman, had brought over some planks of mahogany as ballast, and he thought that the wood might be used up in his brother's building, but the carpenters found the wood too hard for their tools, and objected to use it. Mrs. Gibbons shortly afterwards wanted a small box made, so the doctor called upon his cabinet-maker, and ordered him to make a box out of some wood that lay in his garden. The cabinet-maker also complained that the wood was too hard. But the doctor insisted upon its being used, as he wished to preserve it as a memento of his brother. When the box was completed, its fine colour and polish attracted much attention; and he, therefore, ordered a bureau to be made of it. This was done, and it presented so fine an appearance that the cabinet-maker invited numerous persons to see it, before it was sent home. Among the visitors was her Grace the Duchess of Buckingham, who immediately begged some of the wood from Mr. Gibbons, and employed the cabinet-maker to make her a bureau also. Mahogany from this time became a fashionable wood, and the cabinet maker, who at first objected to use it, made a great success by its introduction

1191. *What is rosewood?*

Rosewood is the wood of a tree which grows in Brazil. It is generally speaking, too dark for large articles of furniture, but is admirably adapted for smaller ones. It is expensive, and the hardness of the wood renders the cost of making articles of it very high

"I am come up to the height of the mountains, to the sides of Lebanon, and will cut down the tall cedars thereof, and the choice fir trees thereof"—
II. KINGS XXIII.

Respecting the other woods used in the manufacture of furniture, we have nothing special to say, except of the oak—the emblem of our native land. This tree yields a most useful and durable wood, and as it not only defends our country by supplying our "wooden walls," but gives to us the floors of our houses, furnishes our good substantial tables, and comfortable arm-chairs, it will be well for us to know a few facts about this celebrated tree. It is said that there are no less than one hundred and fifty species of the oak. The importance of the growth of oaks may be gathered from the fact, that the building of a 70-gun ship would take forty acres of timber. The building of a 70-gun ship is estimated to cost about £70,000. Oak trees attain to the age of 1,000 years. The oak enlarges its circumference from 10½ inches to 12 inches in a year. The interior of a great oak at Allonville, in Normandy, has been converted into a place of worship. An oak at Kiddington, served as a village prison. A large oak at Salcey, was used as a cattle fold; and others have served as tanks, tombs, prisons, and dwelling-houses.

The *Mammoth tree*, which is exhibiting at the Crystal Palace, is one of the great wonders of the vegetable creation. It is the grand monarch of the Californian forest, inhabiting a solitary district on the elevated slopes of the Sierra Nevada, at 5,000 feet above the sea-level. From 80 to 90 trees exist, all within the circuit of a mile, and these varying from 250 to 320 feet in height, and from 10 to 20 feet in diameter. The bark is from 12 to 15 inches in thickness; the branchlets are somewhat pendent, and resemble those of cypress or juniper, and it has the cones of a pine. Of a tree felled in 1853, 21 feet of the bark from the lower part of the trunk were put in the natural form as a room, which would contain a piano, with seats for forty persons; and on one occasion 150 children were admitted. The tree is reputed to have been above 3,000 years old; that is to say, it must have been a little plant when Samson was slaying the Philistines. The portion of the tree exhibiting at the palace is 103 feet in height, and 32 feet in diameter at the base.

1192. *What is tea?*

Tea is the leaf of a shrub (*Thea Chinensis*). The plant usually grows to the height of from three to six feet, and resembles in appearance the well-known myrtle. It bears a blossom not unlike that of the common dog-rose. The climate most congenial to it is that between the 25th and 33rd degrees of latitude. The growth of good tea prevails chiefly in China, and is confined to a few provinces. The *green* and *black* teas are mere varieties, depending upon the culture, time of gathering, mode of drying, &c. *Coffee was used in this country before tea.* In 1664, it is recorded, the East India Company bought 2lb. 2oz. of coffee as a present for the king. In the year 1832, there were 101,687 licensed tea dealers in the United Kingdom. Green tea was first used in 1715. A dispute with America about the duty upon tea led to the American war, out of which arose American independence. The consumption of tea throughout the whole world is estimated at above 52,000,000 lbs., of which the consumption of Great Britain alone amounts to 30,000,000. (*See* 1225).

1193. *What is coffee?*

Coffee is the berry of the coffee plant, which was a native of that part of Arabia called Yemen, but it is now extensively cultivated in India, Java, the West Indies, Brazil, &c. (*See* 1224).

The first coffee-house in London was opened in 1652, under the following circumstances. A Turkey merchant named Edwards, having brought along with him from the Levant, some bags of coffee, and a Greek servant who was skilful in making it, his house was thronged with visitors to see and taste this new beverage. Being desirous to gratify his friends without putting himself to inconvenience, he allowed his servant to open a coffee-house, and to sell coffee publicly.

Here we have another illustration of the great results springing from trifling causes. Coffee soon became so extensively used that taxes were imposed upon it. In 1660 a duty of 4d. a gallon was imposed upon all coffee made and sold. Before 1732 the duty upon coffee was 2s. a pound; it was afterwards reduced to 1s. 6d., at which it yielded to the revenue, for many years, £10,000 per annum. The duty has been gradually reduced, and the consumption has gone on increasing, until at last above 25,000,000 of pounds are consumed annually! Fancy this great result springing from a "friendly coffee party" that assembled in the year 1652.

1194. *What is chocolate?*

It is a cake prepared from the cocoa-nut. The nut is first roasted like coffee, then it is reduced to powder and mixed with water, the paste is then put into moulds and hardened. The properties are very healthful, but its consumption is very insignificant, as compared with tea or coffee. The cocoa tree grows chiefly in the West Indies and South America.

1195. *What is cocoa?*

Cocoa is also a preparation from the seeds or beans of the cocoa tree. But the best form of cocoa for family use is to obtain the beans pure, as they are now commonly sold ready for use, and to break them and then grind them in a large coffee mill.

1196. *What is chicory?*

Chicory is the root of the common endive, dried and roasted as coffee, for which it is used as a substitute. Some persons prefer the flavour of chicory admixed with coffee. But very opposite opinions prevail respecting the qualities of chicory. We belive it to be perfectly healthful, and attribute the prejudice that prevails against it, to its having been used, from its cheapness, to adulterate coffee.

1197. *What is sugar?*

Sugar is a sweet granulated substance, which may be derived from many vegetable substances, but the chief source of which is the sugar cane. The other chief sources that supply it are the maple, beet-root, birch, parsnip, &c. It is extensively used all over the world. Sugar is supposed to have been known to the ancient Jews. It was found in the East Indies by Newcheus, Admiral of Alexander, 325 B.C. It was brought into Europe from Asia.

The art of sugar refining was first practised in England, in 1659, and sugar was first taxed by name by James II., 1685. Sugar is derived from the West Indies, Brazil, Surinam, Java, Mauritius, Bengal, Siam, the Isle de Bourbon, &c. &c. Before the introduction of sugar to this country, honey was the chief substance employed in making sweet dishes; and long after the introduction of sugar it was used only in the houses of the rich. The consumption in England in 1700 reached only 10,000 tons; in 1834 it had reached 180,000 tons. The English took possession of the West Indies in 1672, and in 1646 began to export sugar. In 1676 it is recorded that 400 vessels, averaging 150 tons, were employed in the sugar trade of Barbadoes. Jamaica was discovered by Columbus, and was occupied by the Spaniards, from whom it was taken by Cromwell, in 1656, and has since continued in our own possession. When it was conquered there were only three sugar plantations upon it. But they rapidly increased. Until the abolition of slavery in the West Indies, the production of sugar was almost exclusively limited to slave labour. (*See* 1226).

1198. *What is wheat?*

Wheat, rye, barley, oats, millet, and maize, all belong to the natural order of grain-bearing plants. They all grow in a similar manner, and all yield starch, gluten, and a certain amount of phosphates. They are commonly spoken of as *farinaceous foods.*

From the Sacred writings we learn that unleavened bread was common in the days of Abraham. In the earlier periods of our own history, people had no other method of making bread than by roasting corn, and beating it in mortars, then wetting it into a kind of coarse cake. In 1596, rye bread and oatmeal formed a considerable part of the diet of servants, even in great families. In the time of Charles the First, barley bread was the chief food of the people. In many parts of England it was more the custom to make bread at home then at present. In 1804, there was not a single public baker in Manchester. In France, when the use of yeast was first introduced, it was deemed by the faculty of medicine to be so injurious to health that its use was prohibited under the severest penalties

"I clothed thee also with broidered work, and shod thee with badgers' skin, and I girded thee about with fine linen, and I covered thee with silk."—
EZEKIEL XVI.

Herault says that, during the siege of Paris by Henry the Fourth, a famine raged, and bread sold at a crown a pound. When this was consumed, the dried bones from the charnel house of the Holy Innocents were exhumed, and a kind of bread made therefrom. Bread-street, in London, was once a bread market. From the year 1266, it had been customary to regulate by law the price of bread in proportion to the price of wheat or flour at the time. This was called the assize of bread; but, in 1815, it was abolished. In the year 272 there was a famine in Britain so severe that people ate the bark of trees; forty thousand persons perished by famine in England in 310! In the year 450 there was a famine in Italy so dreadful that people eat their own children. A famine, commencing in England, Wales, and Scotland, in 954, lasted four years. A famine in England and France, in 1193, led to a pestilential fever, which lasted until 1195. In 1315 there was again a dreadful famine in England, during which people devoured the flesh of horses, dogs, cats, and vermin! In the year 1775, 16,000 people died of famine in the Cape de Verds. These are only a few of the remarkable famines that have occurred in the course of history. Let us thank God that we live in times of abundance, when improved cultivation, the pursuit of industry, and the settlement of the laws, render such a calamity as a famine almost an impossibility.

1199. *What is cotton?*

Cotton is a species of vegetable wool, produced by the cotton shrub, called, botanically, *Gossypium herbaceum*, of which there are numerous varieties. It grows naturally in Asia, Africa, and America, and is cultivated largely for purposes of commerce.

The precise time when the cotton manufacture was introduced into England is unknown; but probably it was not before the 17th century. Since then, what wonderful advances have been made! The cotton trade and manufacture have become a vast source of British industry, and of commerce between nations. It was some years ago calculated that the cotton manufacture yielded to Great Britain one thousand millions sterling. The names of Hargreaves, Arkwright, Crompton, Cartwright, and others, have become immortalised by their inventions for the improvement of the manufacture of cotton fabrics. Little more than half a century has passed since the British cotton manufactory was in its infancy—now it engages many millions of capital—keeps millions of work people employed; freights thousands of ships that are ever crossing and re-crossing the seas; and binds nations together in ties of mutual interest. The present yearly value of cotton manufactures in Great Britain is estimated at £34,000,000. About £6,014,000 of the above sum is distributed yearly among working people as wages.

1200 *What is silk?*

Silk, though not directly a vegetable product, is, nevertheless, indirectly derived from the vegetable creation, since it is a thread spun by the silk-worm from matter which the worm derives from the *mulberry leaf*.

Silk is supplied by various parts of the world, including China, the East

"And there was a man in Maon, whose possessions were in Carmel; and the man was very great, and he had three thousand sheep, and a thousand goats: and he was shearing his sheep in Carmel."—I SAMUEL XXV.

Indies, Turkey, &c., where the silk-worm has been found to thrive. The attempts that have been hitherto made to cultivate it in this country have proved unsuccessful. At Rome, in the time of Tiberius, a law passed the senate which, as well as prohibiting the wearing of massive gold jewels, also forbade the men to debase themselves by wearing silk. There was a time when silk was of the same value as gold—weight for weight—and it was thought to grow upon trees. It is recorded that silk mantles were worn by some noble ladies at a ball at Kenilworth Castle, 1286. It was first manufactured in England in 1604. In the reign of Elizabeth, the manufacture of silk in England made rapid strides. In 1666, there were 40,000 persons engaged in the silk trade. The silk throwsters of the metropolis were enrolled in a fellowship in 1562, and were incorporated in 1629. In 1685, a considerable impetus was given to the English silk manufactures. Louis the Fourteenth of France revoked the edict of Nantes. The edict of Nantes was promulgated by Henry the Fourth of France in 1598. It gave to the Protestants of France the free exercise of their religion. Louis the Fourteenth revoked this edict in 1685, and thereby drove the Protestants as refugees to England, Holland, and parts of Germany, where they established various manufactures. Many of these French refugees settled in Spitalfields, and there founded extensive manufactories, which soon rivalled those of their own country; and thus the intolerance of the king was justly punished. What important facts we see connected with the simple thread of the silk-worm!

1201. *What is wool?*

Wool is a kind of soft hair or coarse down, produced by various animals, but chiefly by sheep.

This is another of the useful productions of nature, for which we are indirectly indebted to the vegetable kingdom; for were it not for the rich pastures forming the green carpet of the earth, it would be impossible for man to keep large flocks of sheep for the production of wool. Wool, like the hair of most animals, completes its growth in a year, and then exhibits a tendency to fall off. For the production of wool in England and Wales it has been estimated that there are no less than 27,000,000 sheep and lambs; and, in Great Britain and Ireland, the total number is estimated at 32,000,000. Wool was not manufactured in any quantity in England until 1331, when the weaving of it was introduced by John Kempe and other artizans from Flanders. The exportation or non-exportation of wool has from time to time formed a vexed subject for legislators. Woollen clothes were made an article of commerce in the reign of Julius Cæsar. They were made in England prior to 1200. Blankets were first made in England in 1340. The art of dyeing wools was first introduced into England in 1608. The annual value of the raw material in wool is set down at £6,000,000; the wages of workmen engaged in the wool trade, £9,000,000. The number of people employed is said to be 500,000.

1202. *What is starch?*

Starch is one of the most useful products of the vegetable kingdom. As a rule, *a vegetable, if nutritious at all,* is so *according to the amount of starch which it contains.* It is most abundantly found in the seeds of plants, and especially in the *wheat* tribe.

It is also met with in the cellular tissues of plants, and especially in such underground stems as the *potatoe, carrot, turnip, &c.,* and the stems of the *sago-palm* fig, &c. It is also found in the *bark* of some trees.

1203. *Why is the horse chesnut, though containing a great quantity of starch, unfit for food?*

Because (like many other vegetable productions) it contains with the starch an *acrid juice,* which renders it unhealthy; and although the juice can be separated from the starch, the process is too expensive to be made generally available.

The starch which is used for domestic purposes is an artificial preparation, and does not properly represent the starch of nutrition. A better idea of it is afforded by *the meal of a flowery potatoe.* The starch used by laundresses is frequently prepared from diseased potatoes. This does not impair the quality of the starch, for the purposes of the laundress, and the reason why potatoes that are diseased are thus applied is, that it is one method of saving some part of their value. The finest kinds of starch are prepared from rice. It is prepared by breaking the pulp, and disengaging the starch from the cells; and it is then put through other processes to remove the fragments of the broken cells. But in the flowery meal of the potatoe, the starch cell may be seen entire.

CHAPTER LXII.

1204. *What are vegetable oils and fats?*

Vegetable oils and fats constitute, next to starch and sugar, the most important secretion of the vegetable creation. There are very few plants from which some amount of oil cannot be obtained; and those which are famed for yielding it owe their celebrity rather to the abundance that they yield, and the peculiar qualities of their oil, than to the secretion of oil being rare—for probably there is no plant without it.

Oil is most commonly found in seeds, as *rape-seed, linseed, &c.,* but it is found also in leaves, as in the rose, sweet-briar, peppermint &c., where its presence may be recognised by the distinguishing

"Ointment and perfume rejoice the heart; so doth the sweetness of a man's friend by hearty counsel."—PROVERBS XXVII.

perfume; and it is also found in the wood of a few trees, such as the sassafras and the sandal-wood; the bark frequently yields an oily secretion.

The London and North Western Railway Company alone use about 50,000 gallons of oil yearly.

1205. *Why are fat and oil found most abundantly in the bodies of animals in cold climates?*

Because they contribute to keep the *bodies of animals warm,* not only by their non-conducting property *keeping in* the heat of the animals, but by supplying *carbon* abundantly to combine with *oxygen* during respiration, and thereby developing *animal heat.*

1206. *Why are oil and fat-forming trees found most abundantly in hot climates?*

Because, in hot countries, the formation of large quantities of fat in animal bodies would oppress living creatures with heat; fats and oils are, therefore, produced in those countries chiefly by vegetables, and are used externally by the Asiatics and Africans as an *external* unction for *cooling the skin,* and as *perfumes* which give inspiriting properties to the air, rendered oppressive by excess ot heat.

1207. *Why are succulent fruits most abundant in tropical climates?*

Because they are rendered necessary in those climates by the *excessive heat,* and are found to have a most beneficial effect in cooling, purifying the blood of the inhabitants of tropical countries; while the grandeur of their foliage, and the richness of their flowers, are in perfect keeping with the intensity of light and heat, and serve, by throwing dense shades over the earth, to cool its surface, and to offer to living creatures a pleasant retreat from the rays of the burning sun.

The following sketch of *Botanical Geography* should be read attentively after the reader has gone through the whole of the Chapters of "Reasons." The technical terms employed in the course of the article are nearly all explained at 1212, and should be committed to memory at the commencement of the perusal. *Mimosa* means a sensitive plant; *concentric zones,* circular lines spreading from a centre; *arborescent,* resembling trees; *Gramineæ,* grass-like. The botanical names represent individual plants.

1208. When treating of the geographical distribution of vegetables, we have to mark the general arrangements indicated, and the agencies that have evidently operated in promoting the diffusion of floral tribes. Vegetation occurs over the whole globe, therefore, under the most opposite conditions. Plants flourish in the bosom of the ocean as well as on land, under the extremes of cold and heat in polar and equatorial regions, on the hardest rocks and the soft alluvium of the plains, amidst the perpetual snow of lofty mountains, and in springs at the temperature of boiling water, in situations never penetrated by the solar rays, as the dark vaults of caverns, and the walls of mines, as well as freely exposed to the influences of light and air. But these diverse circumstances have different species and genera. There is only one state which seems fatal to the existence of vegetable life—the entire absence of humidity.

1209. By species we understand so many individuals as intimately resemble each other in appearance and properties, and agree in all their permanent characters, which are founded in the immutable laws of creation. An established species may frequently exhibit new varieties, depending upon local and accidental causes, but these are imperfectly, or for a limited time, if at all, perpetuated.

1210. A genus comprises one or more species similar to each other, but essentially differing in formation, nature, and in many adventitious qualities from other plants. A tribe, family, group, or order, comprises several genera.

1211. The known number of species in the vegetable kingdom has been gradually enlarged by the progress of maritime and inland discovery; but owing to great districts of the globe not having yet been explored by the botanist, the interior of Africa, and Australia, with sections of America, Asia, and Oceanica, it is impossible to state the exact amount. The successive augmentation of the catalogue appears from the numbers below:

	Species.
Theophrastus	500
Pliny	1,000
Greek, Roman, and Arabian botanists	1,400
Bauhin	6,000
Linnæus	8,800
Persoon	27,000
Humboldt and Brown	38,000
De Candolle	56,000
Lindley	86,000
Hinds	89,000

1212. Vegetable forms are divided into three great classes which differ materially in their structure:—1. Cryptogamous plants—those which have no flowers, properly so called, mosses, lichens, fungi, and ferns: as distinguished from those which are phænogamous, or flower-bearing, to which the two following classes belong. 2. Endogenous plants, which have stems increasing from within, also called Monocotyledons, from having only one seed-lobe, as the numerous grasses, lilies, and the palm family. 3. Exogenous plants, which have stems growing by additions from without, also called Dicotelcdons, from the seed consisting of two lobes, the most perfect, beautiful, and numerous class, embracing the forest trees, and most flowering shrubs and herbs.

1213. The exogens furnish examples of gigantic size, and great longevity. In South America on the banks of the Atabapo, Humboldt measured a *Bombax ceiba* more than 120 feet high, and 15 in diameter; and near Cumana, he found the *Zamang del Guayra*, a species of mimosa, the pendant branches of the hemispherical head having a circumference of upwards of 600 feet. The *Adansonia*, or baobab of Senegal, though attaining no great height, rarely more than fifty feet, has a trunk with a diameter sometimes amounting to 34 feet; while the *Pinus Lambertiana*, growing singly on the plains west of the Rocky Mountains, has been found 230 feet high, 60 feet in circumference at the base, 4½ feet in girth at the height of 190 feet, yielding cones 11 inches round, and 16 long. The *Ficus Indicus*, or banian tree, sending out shoots from its horizontal branches, which reaching the ground take root, and form new stems till a single tree multiplies almost to a forest, has been observed covering an area of 1700 square yards.

1214. From the number of concentric zones observed in a transverse section of the stems De Candolle advances proof of the following ages:

Elm	335	years.	
Cypress	about 350	„	
Cheirostemon	„ 400	„	
Ivy	450	„	
Larch	576	„	
Orange	630	„	
Olive	700	„	
Oriental Plane	„ 720	„	and upwards
Cedar of Lebanon	„ 800	„	
Oak	810, 1080, 1500	„	
Lime	1076; 1147	„	
Yew	1214, 1458, 2588, 2880	„	
Taxodium	4000 to 6000	„	
Boabab	5150	„	

1215. Admitting, with Professor Henslow, that De Candolle overrated the ages of these trees one-third, they are examples of extraordinary longevity. Yew trees upwards of 700 years old remain at Fountains Abbey, Yorkshire, as there is historic evidence of their existence in the year 1133. But a yew in the church yard of Darley-in-the-Dale, Derbyshire, is considered by Mr. Bowman as 2000 years old.

1216. The cryptogamous plants afford the most numerous examples of wide diffusion. A lichen indigenous in Cornwall, *sticta aurata*, is also a native of the West India Islands, Brazil, St Helena, and the Cape of Good Hope; while 33 lichens and 28 mosses are common to Great Britain and Australia, though the general vegetation of the two districts is remarkably discordant. Some species of endogenous plants are also widely distributed, the *Phleum alpinum* of Switzerland occurring without the slightest difference at the Strait of Magellan, and the quaking grasses of Europe in the interior of Southern Africa. But only in very few instances are the same species of exogenous plants met with in regions far apart from each other; and generally speaking, in passing from one country to another, we encounter a new flora; for if the same genera occur, the species are not identical, while in districts widely separated the genera are different.

1217. The cryptogamic plants, mosses, lichens, ferns, and fungi, are to the whole mass of phænogamic vegetation in the following proportions in different districts: Equatorial latitudes, 0 deg. to 10 deg.; on the plains, 1-25th, on the mountains, 1-5th; mean latitudes, 45 deg. to 52 deg. ½; high latitudes, 67 deg.

"To give unto them beauty for ashes, the oil of joy for mourning, the garment of praise for the spirit of heaviness; that they might be called Trees of righteousness, The planting of the Lord, that he might be glorified."—ISAIAH LXI.

70 deg., proportion about equal. Thus the proportion of the flowerless vegetation to the flowering increases from the equator to the poles. But the family of ferns *filices*, viewed singly, forms an exception to this law, decreasing as we depart from equinoctial countries, being 1-20th in equatorial and 1-70th in mean latitudes, and not found at all in the high latitudes of the new world.

1218. In equinoctial and tropical countries, where a sufficient supply of moisture combines with the influence of light and heat, vegetation appears in all its magnitude and glory. Its lower orders, mosses, fungi, and confervæ, are very rare. The ferns are aborescent. Reeds ascend to the height of a hundred feet, and rigid grasses rise to forty. The forests are composed of majestic leafy evergreen trees bearing brilliant blossoms, their colours finely contrasting, scarcely any two standing together being of the same species. Enormous creepers climb their trunks; parasitical orchidæ hang in festoons from branch to branch, and augment the floral decoration with scarlet, purple, blue, rose, and golden dyes Of plants used by man for food, or as luxuries, or for medicinal purposes, occurring in this region, rice, bananas, dates, cocoa, cacao, bread-fruit, coffee, tea, sugar, vanilla, Peruvian bark, pepper, cinnamon, cloves, and nutmegs, are either characteristic of it as principally cultivated within its limits, or entirely confined to them.

1219. Rice *(Oryza-sativa)*, the chief food of, perhaps, a third of the human race, is cultivated beyond the tropics, but principally within them, only where there is a plentiful supply of water. It has never been found wild; its native country is unknown; but probably southern Asia.

1220. Bananas, or plantains *(Musa sapientum et paradisiaca)*, are cultivated in intertropical Asia, Africa, and America. The latter species occur in Syria. The banana is not known in an uncultivated state. Its produce is enormous. estimated to be on the same space of ground to that of wheat, as 133 to 1, and to that of potatoes as 44 to 1.

1221. Dates *(Phœnix dactylifera)*, and cocoa *(Cocos nucifera)*, belonging to the family *Palmæ*. The palms, remarkable for their elegant forms and importance to man, contribute more than any other trees to impress upon the vegetation of tropical and equinoctial countries its peculiar physiognomy. The date palm is a native of northern Africa, and is so abundant between the Barbary states and the Sahara, that the district has been named Biledul erid, the land of dates. As the desert is approached, the only objects that break the monotony of the landscape are the date palm, and the tent of the Arab. It accompanies the margin of the mighty desert in all its sinuosities from the shores of the Atlantic to the confines of Persia, and is the only vegetable affording subsistence to man that can grow in such an arid situation. The annual produce of an individual is from 150 to 260lbs. weight of fruit. The cocoa palm furnishes annually about a hundred cocoa-nuts. It is spread throughout the torrid zone; but occurs most abundantly in the islands of the Indian archipelago. The family of palms is supposed to contain a thousand species, some of large size, forming extensive forests.

1222. Cacao *(Theobrama cacao)*, from the seeds of which chocolate is prepared, grows wild in central America, and is also extensively cultivated in Mexico, Guatamala, and on the coast of Cumana.

1223. Bread-fruit tree *(Artocarpus incisa)*, a native of the South Sea Islands, and Indian archipelago, grows also in Southern Asia, and has been introduced

"And they returned and prepared spices and ointments; and rested the Sabbath-day, according to the commandment."—LUKE XXIV.

into the tropical parts of America; but the fruit is not equal to the banana as an article of human food.

1224. Coffee *(Coffea Arabica)*. The bush has probably for its native region the Ethiopian Highlands, from whence it was taken in the fifteenth century to the Highlands of Yemen, the southern part of the Arabian peninsula. It has been introduced, and is now extensively cultivated in British India, Java, Ceylon, the Mauritius, Brazil, and the West Indies, but the quality is inferior, which makes the climate of the Mocha coffee district of importance, as peculiarly favourable to the plant. It grows there on hills described by Niebuhr as being soaked with rain every day from the beginning of June to the end of September, which is carefully collected for the purpose of irrigation during the dry season. Forskhal gives the following temperatures in the district:

Boit el Fakih ...	March 16,	7 A.M. 76 deg	1 P.M 95 deg.	10 P.M. 81 deg.
„ ...	„ 18,	„ 77	„ 95	„ 81
Hodeida	„ 18,	„ 72	„ 92¾	„ 78
Bulgosa, a village in the hills ...	„ 20,	„ 69½	„ 85½	„ 73

1225. Tea *(Thea Chinensis)*. The plant is indigenous in China, Japan, and Upper Assam. In the latter country, it has recently been found in a wild state, and is in process there of extensive cultivation. As the plant is hardy, its culture has very lately been attempted in the South of France, and apparently with complete success. A similar experiment on the burning plains of Algeria completely failed, all the plants being killed by the heat, notwithstanding every precaution. Tea was first introduced into Europe by the Dutch in 1666. The leaves of the coffee-plant have long been used as a substitute for tea, by the lower classes in Java and Sumatra; and recently, Professor Blume, of Leyden, exhibited samples of tea prepared from coffee-leaves, agreeing entirely in appearance, odour, and taste, with the genuine Chinese production.

1226. Sugar-cane *(Saccharum officinaram)*, a species of *Gramineæ*, occurs to some extent without the tropics, having been cultivated centuries ago in Europe, as at present scantily in the South of Spain. But it properly belongs to the torrid zone, and has for its principal districts, the Southern United States, the West Indies, Venezuela, Brazil, the Mauritius, British India, China, the Sunda and Philippine Islands. The plant was found wild in several parts of America on the discovery of that continent, and occurs in a wild state on many of the islands of the Pacific.

1227. Vanilla *(Vanilla aromatica)*, the fruit of which forms the well-known aromatic, grows wild principally in Mexico.

1228. Peruvian bark (*Cinchona officinalis*), a forest tree, of which there are several species, furnishing the valuable medicine so called. It is exclusively confined to South America, and grows chiefly on the Andes of Loxa and Venezuela.

1229. Pepper (*Piper nigrum)* belongs exclusively to the Malabar coast, where it has been found wild, Sumatra, which produces the greatest quantity, Borneo, the Malay peninsula, and Siam. Other species of *Piperaceæ* occur in tropical America.

1230. Cinnamon (*Laurus Cinnamomum)*, a small tree yielding the aromatic bark, is found native only in the island of Ceylon; but another species occurs in Cochin China.

"I am the true vine, and my Father is the husbandman."—JOHN XV.

1231. Clove (*Myrtus caryophyllus*), an evergreen small tree, the dried flower-buds of which form the celebrated aromatic, grows naturally in the Moluccas whence it has been conveyed to other tropical districts. The island of Amboyna, one of that group, is the principal seat of its cultivation. The lowest temperature there is 72 degs.; the mean temperature of the year 82 degs.

1232. Nutmeg *(Myrstica moschata)* grows naturally in several islands of the eastern archipelago, but is principally cultivated in the Banda Isles.

Tropical families and forms successively vanish with an increase of distance from the equator, and new phases of vegetation mark the transition from hot to temperate climates. Vividly green meadows, abounding with tender herbs, replace the tall rigid grasses which form the impenetrable jungle; and instead of forests composed of towering evergreen trees, woods of the deciduous class appear, which cast their leaves in winter, and hybernate in the colder season, the oak, ash, elm, maple, beech, lime, alder, birch, and sycamore. The cultivation of the vine becomes characteristic, with the perfection of the cereal grasses, and a larger proportion of herbaceous annuals and cryptogamic plants.

1233. The vine *(Vitis vinifera)* is less impatient of a cold winter than a cool summer. Hence its northern limit, which coincides with lat. 47 deg. 30 min. on the west coast of France, rises in the interior, where, though the winters are colder, the summers are warmer, to lat. 49 degs., cuts the Rhine at Coblentz in lat. 50 deg. 20 min., and ascends to 52 deg. 31 min. in Germany.

1234. Receding further from the equator, magnificent forests of the fir and pine tribe prevail, as in the central parts of Russia, on the southern shores of the Baltic, in Scandinavia, and North America. But some of the cereals are no longer cultivatable, and several timber-trees common to the temperate zone do not reach its northern limits. Gradually all ligneous vegetation disappears entirely as higher latitudes are approached, the woods having first dwindled to mere dwarfs in struggling with the elements, hostile to that state which nature destined them to assume. The limit of the forests is a sinuous line running along the extreme north of the old world; and extending from Hudson's Bay, lat. 60 deg., to the Mackenzie River, lat. 68 deg., and thence to Behring's Strait. The dwarf birch *(Betula nana)*, a mere bush, is the last tree found on drawing near the eternal snow of the pole. At the island of Hammerfest, lat. 70 deg. 40 min., near the North Cape, it rises to about the height of a man, in sheltered hollows between the mountains, its lower branches trailing on the ground, affording a shelter to the ptarmigan. In the polar zone, some low flowering annuals, saxifrages, ranunculi, gentians, chickweeds, and others, flourish during the brief ardent summer; a few perennials also accommodate themselves to the rigorous climate by spreading laterally, never rising higher than four or five inches from the ground; till finally no development of vegetable life is met with, but lichens, and the microscopic forms that colour the snow.

1235. In Europe, wheat ceases with a line connecting Inverness in Scotland, lat. 58 deg., Drontheim in Norway, lat. 64 deg., and Petersburgh in Russia lat. 60 deg. 15 min. Oats reach a somewhat higher latitude. Barley and rye ascend to lat. 70 deg., but require a favourable aspect and season to produce a crop.

1236. The northern limit of the growth of oak, lat. 61 deg., falls short of that of wheat. The oak makes a singular leap at the confines of Europe and Asia, disappearing towards the Ural mountains. This is the case also with the wild-nut and apple. The oak and the wild-nut, however, re-appear suddenly in

"He hath made the earth by his power, he hath established the world by his wisdom, and hath stretched out the heavens by his discretion."—JEREMIAH X

Eastern Asia, on the banks of the Argoun and the Amour; and the apple occurs again in the Aleutian Isles.

1237. The following are the northern limits of several trees in Scandinavia:

	Lat.
Beech, *Fagus silvatica*	60 deg. 0 min.
Hard Oak, *Quercus robur*	61 " 0 "
Common Elm, *Ulmus campestris*	61 " 0 "
Common Lime, *Tilia communis*	61 " 0 "
Common Ash, *Fraxinus excelsior*	62 " 0 "
Fruit trees	63 " 0 "
Hazel, *Corylus avellana*	64 " 0 "
Spruce Fir, *Abies excelsa*	67 " 40 "
Service Tree, *Sorbus aucuparia*	70 " 0 "
Scotch Fir, *Pinus silvestris*	70 " 0 "
White Birch, *Betula alba*	70 " 40 "
Dwarf Birch, *Betula nana*	71 " 0 "

1238. Thus distinct vegetable regions are observed on passing from south to north through different climatic zones, defined as to their limits by the isothermal curves, and not by the parallels of latitude. Similar changes of vegetation mark a perpendicular transit through varying climates. A succession of plants appear on the tropical mountains which rise above the snow line, corresponding to those which are encountered in mean and high latitudes. The higher we ascend, the more does the number of the phænogamic class diminish in proportion to the cryptogamic, till only members of the latter family are found, whose further progress upward is arrested by the everlasting snow. The last lichen met with by Saussure on Mont Blanc, *Silene acaulis*, was also observed by M. Brevais in the neighbourhood of Bosekop, lat. 69 deg. 58 min. where it was vegetating on the sea-shore, shaded by the last pines of Europe.

1239. Isolated mountains display to the best advantage the effect of climatic change of vegetation.

1240. Etna is divided into three great regions: *La Regione Culta*, or fertile region; *La Regione Sylvosa*, or woody region; *La Regione Deserta*, the bare or desert region. But each of these is susceptible of sub-divisions, defined by the presence of certain families of plants, forming seven botanical zones.

1. The sub-tropical zone, which does not rise more than 100 feet above the evel of the sea, is characterised by the palm, banana, Indian fig, sugar-cane, varities of mimosa and acacia, which with us are only found in conservatories.

2. The hilly zone rises about 2,000 feet, characterised by the orange, lemon, shaddock, maize, cotton, and grape plants.

3. The woody zone lies between the height of 2,000 and 4,000 feet, where the cork-tree flourishes, several kinds of oak, the maple, and enormous chesnuts.

4. The zone between the height of 4,000 and 6,000 feet is distinguished by the beech, Scotch fir, birch, and, among small plants, by clover, sandwort, chickweed. dock, and plantain.

5. The sub-alpine zone, between the elevation of 6,000 and 7,500 feet, produces the barberry, soap-wort, toad-flax, and juniper.

6. The zone between 7,500 and 9,000 feet, has almost all the plants of the preceding, with the fleshy and jagged groundsel.

"In the mountain of the height of Israel will I plant it; and it shall bring forth boughs, and bear fruit, and be a goodly cedar: and under it shall dwell all fowl of every wing; in the shadow of the branches thereof shall they dwell."—
EZEKIEL XVII.

7. The narrow zone between 9,000 and 9,200 feet, only produces a few lichens, beyond which, there is complete sterility.

1241. The Peak of Teneriffe exhibits five botanical districts, thus distinguished by Von Buch:

1. The region of Africa forms, 0—1,248 feet, comprising palms, bananas, the sugar-cane, various species of arborescent *Euphorbiæ*, *Mesembryanthema*, the *Dracæna*, and other plants, whose naked and tortuous trunks, succulent leaves, and bluish-green tints, are distinctive of the vegetation of Africa.

2. Region of Vines and Cereals, 1,248—2,748 feet, comprising also the olive, and the fruit-trees of Europe.

3. Region of Laurels, 2,748—4,350 feet, including lauri of four species, the wild olive, an oak, the iron-tree, the arbutus, and other evergreens. The ivy of the Canaries and various twining shrubs cover the trunks of the trees, and numerous species of fern occur, with beautiful flowering plants.

4. Region of the Pines, 4,350—6,270, characterised by a vast forest of trees resembling the Scotch fur, intermixed with juniper.

5. Region of the Retama, 6,270—11,061 feet, a species of broom, which forms oases in the midst of a desert of ashes, ornamented with fragrant flowers, and furnishing food to the goats which run wild on the Peak. A few gramineous and cryptogamic plants are observed higher, but the summit is entirely destitute of vegetation.

1242. There are many plants which can accommodate themselves to the most diverse climates and localities; and therefore ascend from the plains close to the boundary of vegetable life on the highest mountains. But it is the general law in these cases for such plants to be singularly modified in appearance and anatomical structure as they ascend. The spring gentian, *Gentiana verna*, is one of the exceptions, which Raymond found unaltered at all heights in the Pyrenees.

1243. Trees, plants, and bushes, of humbler growth, which occur on the plains and at great heights, are usually much smaller in the latter situation. The leaves, and everything green about them, dwindle with the increased elevation; and the pure, well-defined green is exchanged for an ill-defined light yellow. Singular enough, those parts which seem most capable of resisting cold, as the leaves and stalks, are uniformly subjected to a diminution of their vital functions; while the flowers remain of the same size, are never deformed, and become more dense and richer in their colours. While the *Myosotis silvestris* becomes stunted, its flowers assume an intense blue—the admiration of the traveller. The flowers of the pale primrose have a much deeper colour on the top of the Faulhorn, while the plant itself is much smaller than its congener on the Swiss plains. The observations of M. Parrot, among others, are to this effect on the flora of the Caucasus, of Ararat, the Swiss and Italian Alps, and the Pyrenees. The arctic flora is similarly distinguished.

1244. The preceding references to different climatic states are, however, perfectly inadequate to explain the phenomena of vegetable distribution. While an analogy is often observable between the plants of different regions under corresponding circumstances of latitude, elevation, and soil, the species are generally found to be different; and usually the botanical character of countries not widely apart from each other, is totally different, though under the same parallels.

1245. Some plants are entirely confined to one side of our planet. The beautiful genus *Erica*, or heath, of which there are upwards of 300 species, occurs with breaks over a narrow surface, extending from a high northern latitude to the Cape of Good Hope. But the whole continent of America does not contain a single native specimen; nor has a *Pænia* been found in it, except a solitary one to the west of the Rocky Mountains. On the other hand, the New World contains many families, as the *Cacti*, which are not found naturally in the Old.

1246. Some plants occur in a single specific locality, frequently a contracted area, and nowhere else. The beautiful *Disa grandiflora* is limited to a spot on the top of the Table Mountain at the Cape; and the celebrated cedar of Lebanon appears to be restricted in its spontaneous growth to the Syrian mountains. The small island of St. Helena has an indigenous flora, with a few exceptions different from that of the rest of the globe.

1247. Mountain chains of no great width very commonly divide a totally distinct botany. There is a marked difference in the vegetation of the Chilian and opposite side of the Andes, though the climate as well as the soil is nearly the same, and the difference of longitude very trifling. In North America, two completely different classes of vegetation appear on the two sides of the Rocky Mountains. A variety of oaks, palms, magnolias, azaleas, and magnificent rhododendrons occur on the eastern side, all of which are unkn own on the western, the region of the giant pine.

1248. The distinct vegetation possessed by various parts of the globe, has led to its division into botanical kingdoms or phyto-geographical regions, named in general after the genera that are either peculiar to them, or predominant in them. The arrangement of M. Schouw which is usually adopted, discriminates twenty-five great provinces of characteristic vegetation upon the surface of the earth.

In constituting any portion of the globe into a phyto-geographical region, M. Schouw has proceeded upon the following principles:—1. That at least one-half of the species should be indigenous in it. 2. That a-quarter of the genera should also be peculiar to it, or at least should have a decided maximum. 3. That individual families of plants should either be exclusively confined to the region, or have their maxima there.

1249. The phenomena of botanical geography, and the facts of geology, are mutually illustrative. The existing dry land having been upheaved above the waters at different epochs, it may be reasonably inferred that each portion on its emergence received a vegetable creation in harmony with its position. The ultimate constitution of the general surface into different botanical kingdoms would hence follow, each of which has preserved its primitive features, while adjoining, and even far distant foci, have to some extent intermingled their respective products, under control of the natural agencies of diffusion.

1250. The agents that involuntarily officiate in the diffusion of vegetable products are the atmosphere, the waters, and many animals.

1. The impulsion of the atmosphere in its calmest state, is quite sufficient to transport to considerable distances seeds furnished with downy appendages or winglets, as is the case with many plants, with the minute sporules of

cryptogamia, which are light as the finest powder. When ordinary breezes convey the sand-dust of the Sahara a thousand miles or more from the desert, it may be conceived that seeds, which are comparatively heavy, are borne far from home by the hurricane. Two Jamaica lichens, which had never been seen in France before, were found by De Candolle growing on the coast of Brittany the offspring of sporules which had been swept over the Atlantic.

2. The mountain torrent washes down into the valley the seeds that have accidentally fallen into it, or have been swept away by its overflows; and hence the plants of the High Alps occur on the plains of Switzerland, which are entirely wanting in France and Germany. Rivers answer the same purpose more extensively, and also the oceanic currents. The nicker-tree, one of the leguminous tribe, has been raised from seed borne across the Atlantic by the Gulf stream.

3. Animals of the sheep and goat kinds, with the horse, deer, buffalo, and others, widely disperse several species of plants, the seeds of which, furnished with an apparatus of barbs and hooks, adhere to their coating. Seeds also of various kinds pass through the digestive organs of birds, uninjured as to their vitality. The little squirrel buries the acorn in the ground for winter provender, and sows an oak, if prevented from returning to the spot.

1251. Plants capable of extended naturalisation, and serviceable as articles of food or luxury, have been widely disseminated by the human race in their migrations. The cerealia afford a striking example. These important grasses known to the ancients, wheat, barley, oats, and rye, were the gifts of the Old World to the New. They are also importations into Europe; but the loose reports of the ancients, and the diligent researches of the moderns, alike leave us in ignorance of their native seat. Probability points to the conclusion that they have spread from the neighbourhood of the great rivers of Western Asia, the primitive location of the human family; and it is not impossible that in that imperfectly explored district, or further east on the Tartarian table-land, some of the cereals may yet be found growing spontaneously. The first wheat sown in North America, consisted of a few grains accidentally found by a negro slave of Cortes, among the rice taken for the support of his army. In South America the first wheat was brought to Lima by one of the early colonists, a Spanish lady, Maria d'Escobar. An ecclesiastic, Jose Rixi, was the first to sow it in the neighbourhood of Quito.

1252. Maize, or Indian corn *(Zea mays)*, has been dispersed in the Old World from the New; and also a more important product, the potato (*Solanum tuberosum)*, the use of which now extends from the extremity of Africa to Lapland. In Chili, the native country of the plant, it occurs at present in a wild state. The Spaniards imported it into Spain, and from thence it was communicated to Italy. It was first made known in England at a subsequent period from Virginia, having been received there from the Spanish colonists in South America, as it is not a native of intervening Mexico.

1253. The grape-vine, so extensively spread over Europe, is probably not indigenous in any part of it. It chiefly owes its diffusion there to the Romans, who received it from the Greeks, to whom it most likely immediately came from the country between the Black and Caspian Seas. The Romans introduced most of the finer European fruit-trees, some from Africa, as the pomegranate, but the great majority from Western Asia, as the orange, fig, cherry, peach, apricot, apple, and pear. A variety of the plum, the damson, or damascene, came from the neighbourhood of Damascus during the Crusades The name of the damask-rose points to the importation of the plant from the same quarter into Europe.

The ocean as well as the land has different botanical regions; and changes of the vegetation are observed with the depth analogous to the variations of terrestrial plants with the height. Marine vegetation seems to have its vertical extent determined by the range of light in water, which varies with the power of the sun and the transparency of the water.

CHAPTER LXIII.

1254. *What are vegetable gums?*

Vegetable gums are secretions of plants which are generally *soluble in water*, and which subserve various useful purposes. *Gum Arabic* is one of the most important of this class of vegetable productions.

Gutta-percha is an invaluable substance lately added to the list of known vegetable productions. It is obtained by cutting the bark of trees of the class called *Sapotacea*. Its proper name is gutta Pulo Percha, gutta meaning gum, and Pulo Percha is the island whence it is obtained. But gutta-percha is not, strictly speaking, a gum.

India-rubber is also a vegetable secretion, improperly called elastic gum. It is obtained from the milky juice of various trees and plants, especially from the syringe tree, of Cayenne.

1255. *What are vegetable resins?*

Vegetable resins are derived from the secretions of plants, and are generally distinguished from gums by being *insoluble in water*, but being soluble in spirits.

When one of these substances is soluble in either water or spirits it is called a *gum-resin*.

1256. *What are vegetable acids?*

Vegetable acids are chiefly obtained from *fruit*; but also abundantly from *wood*, by distillation.

"Thou art the God that doest wonders."—PSALM LXXVII.

1257. *What is tannin?*

Tannin is a vegetable production, obtained chiefly from the oak-bark, and from a variety of other vegetable sources. It possesses the peculiar chemical property which renders it valuable in tanning leather.

1258. *What is opium?*

Opium is the produce of the *poppy*, and is obtained from the seed.

1259. *What are vegetable dyes?*

Vegetable dyes are the various colours derived from the secretions of plants, such as *indigo, madder, logwood, alkanet-root, &c.*

1260. *What is silica?*

Silica is a mineral substance, commonly known as *flint;* and it is one of the wonders of the vegetable tribes, that, although flint is so indestructible that the strongest chemical aid is required for its solution, plants possess the power of *dissolving and secreting* it. Even so delicate a structure as the wheat straw dissolves silica, and every stalk of wheat is covered with a perfect, but inconceivably thin coating of this substance.

Amid all the wonders of nature which we have had occasion to explain, there is none more startling than that which reveals to our knowledge the fact that a flint stone consists of the mineralised bodies of animals, just as coal consists of masses of mineralised vegetable matter. The animals are believed to have been infusorial animalculæ, coated with silicous shells, as the wheat straw of to-day is clothed with a glassy covering of silica. The skeletons of animalculæ which compose flint may be brought under microscopic examination. Geologists have some difficulty in determining their opinions respecting the relation which these animalculæ bear to the flint stones in which they are found. Whether the animalculæ, in dense masses, form the flint; or whether the flint merely supplies a sepulchre to the countless millions of creatures that, ages ago, enjoyed each a separate and conscious existence, is a problem that may never be solved. And what a problem! The buried plant being disentombed, after having lain for ages in the bowels of the earth, gives us light and warmth; and the animalcule, after a sleep of ages, dissolves into the sap of a plant, and wraps the coat it wore, probably "in the beginning, when God created the heavens and the earth, and when the earth first brought forth living creatures," around the slender stalk of waving corn!

1261. *Why is silica diffused over the stems of wheat, grasses, canes, &c.?*

Because it affords strength, density, and durability, to structures that are very light, and which, but for this beautiful provision, would be exceedingly perishable.

1262. *Why is guano a productive manure?*

Because it contains, with other suitable elements, an abundance of the *silicous skeletons of animalculæ.*

1263. *Why does a wheat-crop greatly exhaust the soil?*

Because, as well as the *carbon, and the salts,* which form the straw and the grain, it draws off from the soil a great amount of *silica.*

1264. *Why is straw frequently used as a manure?*

Because it gives back, with other substances, a *considerable proportion of silica,* in that form which adapts it to the use of the succeeding crop.

1265. *Why is the structure of herbaceous plants less consolidated than that of woody plants?*

Because, for the most part, herbaceous plants last only *a single year;* they, therefore, do not require the enduring qualities of plants that have to sustain the influences of the elements for a succession of seasons.

1266. *Why are the stalks of plants of light structure generally cylindrical?*

Because the cylindrical form is stronger than any other; *a hollow cylinder,* with moderately thick walls, *is stronger than a solid rod,* containing the same amount of material.

1267. *Why do the stalks of plants become hollow?*

Because the parallel and perpendicular fibres of the stalk are developed *more rapidly than the horizontal.* The growth of the plant therefore, consists of a kind of *divergence from the centre.*

1268 *Why are the stomata, or pores of leaves, generally placed on their under surface?*

Because, being placed on the under surface, they are *shaded* from the action of the *sun's rays,* and so carry on the function of respiration more actively than if subjected to direct heat; they are also protected from the injurious *effects of dust;* and are moistened by *evaporation from the earth's surface.*

1269. *Why have plants a formation of pith in their centre?*

The pith is the chief organ of nutriment, especially in the young plant. It is the structure which first conveys fluids to, and receives them from, the newly-formed leaf. It communicates with every branch, leaf, bud, and flower; and also with the bark, through the *medullary rays,* which radiate horizontally from the centre of the plant. It is the centre of the movements of the sap which occur in the horizontal vessels; and it holds an important influence over the life of the plant.

1270. *Why are trees covered with bark?*

Because the bark serves to protect the woody structure, and also to give a passage to the descending sap which flows abundantly in the spring, and out of which the woody fibre is formed. It is also, from its peculiar nature, well fitted to endure the changes of the seasons for many years; and from its non-conducting properties it serves to maintain the equal temperature of the vital parts of the tree.

1271. *What is cork?*

Cork is the bark of a description of *oak-tree,* which grows in great abundance in Spain, Italy, and France.

1272. *Why does the cork-tree release its own bark?*

Because it possesses a bark which is exceedingly *useful to man;* and it seems, therefore, to have been the design of providence that the tree should cast it off, to be applied to the wants of the human family; for the cork-tree does not discharge its bark by the mere cracking, or exfoliation, of its substance; the tree retains the bark for a number of years, until it has attained that consistency and thickness which renders it useful, and then the tree forms within the bark a series of tabular cells, which *cut off the connection of the bark with the internal structure,* after which it peels off in large sheets.

"And all the trees of the field shall know that I the Lord have brought down the high tree, have exalted the low tree, have dried up the green tree, and have made the dry tree to flourish: I the Lord have spoken, and have done it.—EZEK XVII.

Man assists this evident intention of nature, by slitting the bark from the top of the tree to its base; but even were this not done, the bark would be cast off by the tree itself

Another proof of design in this useful adaptation of the cork-tree is to be found in the fact, that it thrives under treatment that would destroy other trees. The cork-tree will endure the barking process for *seven or eight successive years.*

CHAPTER LXIV.

1273. *Why are there curious markings in walnut, mahogany, rose-wood, satin-wood, &c.?*

Because those markings are produced by the various *structure of the vessels* by which the wood is formed; and by successive zones of wood, which indicate the periods of growth.

The inclosure of zone within zone is owing to the mode in which the wood is produced, and the position in which it is deposited. Wood is formed by the leaves during the growing season, and passes down towards the root between the bark and the wood of the previous year (if any), or in the position in which cambium is effused; and, as the leaves more or less surround the whole stem, the new layer at length completes a zone, and perfectly encloses the wood of all former years. This is the explanation of the term *exogenous*, which is derived from two words signifying to grow outwardly, for the stem increases in thickness by successive layers on the outer side of the previously-formed wood. That this is the mode of growth has been abundantly proved by experiment, and demonstrated by accidental discoveries. Thus, if a plate of metal be inserted between the bark and wood, it will, in progress of time, become inclosed by the new wood which has overlaid them. So in like manner if letters be cut deeply through the bark and into the wood, the spaces will not be filled up from the bottom, but may be seen in subsequent years overlaid by new wood. A statement appeared in a daily paper, during the past year, to the effect that in cutting down a tree a cat had been discovered inclosed in the wood of the trunk, These facts prove that the wood is applied from without. Again, if a branch be stripped of its leaves down to a certain point, it will not grow above that point; and so, in like manner, if branches be stripped from one side of a tree, the tree will not grow on that side. If a circle of bark be removed from a branch above and also below a leaf, it will be found that increase of size will occur below, but not above that bud; and so, likewise, whenever a ring of bark is removed from a tree, the new woody fibre will not proceed from the lower but from the upper edge.—*Orr's Circle of the Sciences.*

"And when he saw a fig tree in the way, he came to it, and found nothing thereon, but leaves only, and said unto it, Let no fruit grow on thee henceforward for ever. And presently the tree withered away."—MATTHEW XXI.

1274. *Why have trees with large trunks a great number of leafy branches?*

Because it is *by the leaves* that the secretion is formed which supplies the *woody fibre.* The number of leaves on a tree, therefore, generally bears a relation to the size of its trunk, and the number of its branches.

1275. *Why have poplar-trees comparatively few branches and leaves?*

Because their trunks are comparatively *small*, although they grow to a great height.

1276. *Why had the mammoth-tree comparatively few leaves in relation to the immense size of its bark?*

Because the woody texture of this tree (*Wellingtonea gigantea*) is *exceedingly light and porous.* It is, in fact, lighter than cork, and, therefore, requires less leaf-produce in its formation.

1277. *Why have oak-trees an abundance of leaves?*

Because their wood is *so dense* that they require a larger amount of the wood-forming secretion which is supplied by the leaves.

1278. *Why are the trunks of trees round?*

Because, generally speaking, the leaves are distributed upon branches around the trees in every direction. They consequently send down the wood-forming principle on all sides. When a trunk is unduly developed on one side, it may generally be traced to the unequal distribution of the branches.

1279. *What are exogenous stems?*

Exogenous stems are those that grow by the addition of wood *on their outer surface,* underneath the bark.

1280. *What are endogenous stems?*

Endogenous stems are those that *grow inwardly*, from the centre. Trees of this class, of which palms are the best example, are almost peculiar to tropical climates.

1281. *Why do endogenous stems chiefly abound in tropical climates?*

Because, probably, the excessive heat of those climates would interfere with the *formation of wood from the sap* upon the outer surface.

The vascular structure of endogenous stems lying more abundantly towards their centre, tends to *conserve the juices* which in hot climates are so highly valued. Palm-wine is a delicious and cooling beverage, and is procured from various kinds of palms, but especially from the cocoa-nut palm. Even the fresh sap is very refreshing. The juice is procured by cutting the tree in the upper part, and attaching a vessel to the opening, to receive the sap. Its flow is increased by cutting off a slice of the wood daily.

1282. *Why have endogenous stems no bark?*

Because, one of the chief functions of the bark in exogenous trees, is to *protect the sap* from which the wood is formed on the outward surface; and as there is no such external flow of sap in endogenous trees, the bark is *unnecessary to them*, and is therefore withheld. They are furnished instead with a thin cuticle.

1283. *Why do endogenous stems grow to a great height?*

Because, as the stem grows from the centre, it soon reaches that limit of diameter *which its vascular structure is calculated to support;* and, therefore, the wood-forming sap is deposited chiefly at the top of the stem, causing it to grow to a considerable height.

1284. *Why do the various vegetable fruits ripen in succession?*

Because the Author of Nature has thus arranged its economy, *in order that the wants of living creatures may be adequately provided for.* Some vegetable productions arrive at their perfection in the spring; others in summer; and others in autumn. Among the latter are many that require to come slowly to maturity after they are gathered; by these the winter season is provided for, and a surplus of the winter stock goes to supply the natural deficiency of spring.

1285. *Why, when seeds are sown, and germination begins, does the leaf-germ seek the light, and the root-germ grow down into the earth?*

Because the Creator has endowed every single seed with a

vital instinct which governs its development. The rootlet could more easily grow upward than downward, because of the looser earth, and of the exciting influences of light and moisture. Yet it takes the contrary course, leaving the leaf-germ to come up to meet the sun-light, and to send down to the stem and roots, the matter needed for their growth.

Frequently, indeed, when seeds are thrown into the earth, their natural position is reversed, and when the germs first start from the seed, the *root-germ* is directed *upward* and the *leaf-germ downward.* What then occurs? They each turn, and, in doing so, frequently cross each other. Each goes to its particular duty—the duty that God appointed.

CHAPTER LXV.

1286. *Why are the seeds of plants indigestible?*

Because they are encased in a hard covering upon which the gastric juice of animals takes no effect. This provision has been made by the Creator, *for the preservation of seeds,* the productions of which are so essential to animal life.

The gastric juice can dissolve any other part of the plant, even the woody fibre, and yet upon the *seed* it takes no effect. When, however, the seed is *crushed,* and, thereby, the vital principle destroyed, so that no plant can spring from it, the gastric juice acts upon it, and it is soon dissolved.

Hence graminivorous birds are provided with gizzards *to break the protecting coats of the grain;* and animals that feed on seeds and nuts *strip them of their shells and husks.*

It is remarkable that in the *succulent fruits,* such as the strawberry, the raspberry, currant, apple, orange, melon, &c., and which, from their very nature, are likely to attract animals to use them, and in eating which *the seeds are likely to be swallowed,* they are fortified by a doubly-protective coating; the pips of the apple, orange, &c., and the seeds of the strawberry and raspberry, pass through the digestive organs, not only unharmed, but their

"And it was commanded them that they should not hurt the grass of the earth, neither any green thing, neither any tree."—REVELATION IX.

germinating powers are even improved by the warmth and trituration of the stomach. Indeed, the stomachs of quadrupeds and birds have been made the vehicles of propagating plants, and distributing them to the widest geographical latitudes. It is even said of some seeds that they will not germinate until they have passed through the digestive ogans of an animal.

1287. *Why do animals that graze, crop the tender blades of grass, but avoid the tall stems?*

Because they are tempted by the greater sweetness and tenderness of the young blades; and in this temptation a very important end is served; for, by avoiding the stems that have grown up, *the animals spare the matured plant by which seeds are borne,* and by which the supply of food is to be continued.

1288. *Why do the eggs of butterflies lie dormant during the winter?*

Because the *coldness of the winter* would be fatal to the life of the young insects; and the absence of vegetation would leave the caterpillars to *perish of starvation,* if they were developed during the winter months.

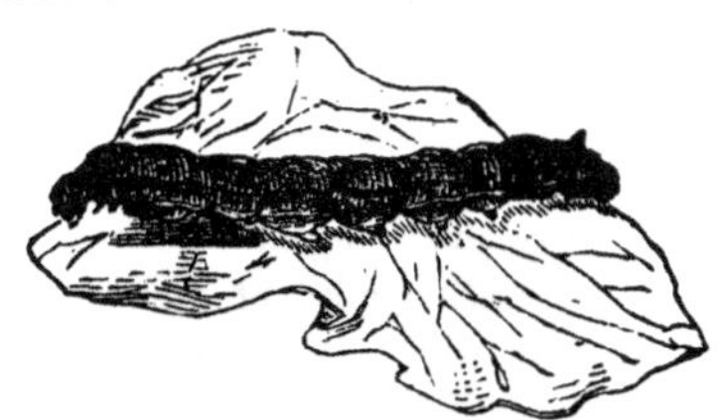

Fig. 76.—CATERPILLAR FEEDING.

1289. *Why do caterpillars appear in the spring?*

Because the increasing warmth of the sun developes the living embryo, *at the same time that it developes the vegetable germ.* The warmth, therefore, that calls the caterpillar from its embryo sleep, also kindles the germinating power of the vegetable upon which it is destined to feed. The worm awakes and finds the bountiful table of nature spread for it.

"Thou shalt plant vineyards, and dress them, but shalt neither drink of the wine, nor gather the grapes: for the worms shall eat them."—
DEUTERONOMY XXVIII.

1290. *Why does the caterpillar eat voraciously?*

Because it *grows rapidly*, and a large amount of vegetable matter is necessary to supply the rapid growth of its animal substance. Caterpillars in the course of a month devour 60,000 times their own weight of aliment.

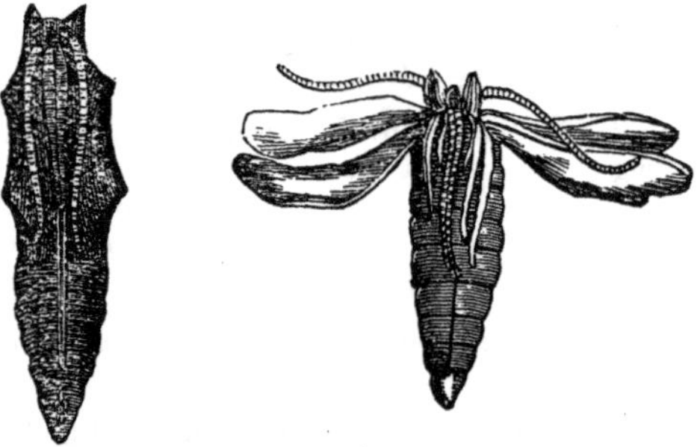

Fig. 77.—THE UNDER SIDE OF THE CHRYSALIS OF THE PEACOCK BUTTERFLY.

Fig. 78.—THE SAME CHRYSALIS, WITH PART OF ITS SHEATH RAISED TO SHOW THE PARTIALLY-FORMED WINGS, &c.

1291. *Why do caterpillars pass into the state of the chrysalis?*

Because they are thereby prepared for the new existence which they are about to enjoy; *new organs must be perfected in them* to adapt them to the altered conditions of their lives.

Because, also, in the transformation of their bodies, differing materially from the laws of existence that pertain to other creatures, the Creator affords another illustration of his Omnipotence.

Because, also, during the stage that the insect sleeps in the chrysalis, the flowers and their sweet juices, upon which the fly is to feed, are being prepared for it, just as, when it was sleeping in the egg, the green food was being prepared for the caterpillar. When, therefore, the beautiful fly spreads its silken wings, it finds a *second time* that, while it has slept, its meal has been prepared, and it now flies away joyously to feed upon the milk and honey of beautiful flowers which, at the time it passed into the chrysalis, had not yet unfolded their petals.

"For the moth shall eat them up like a garment, and the worm shall eat them like wool: but my righteousness shall be for ever, and my salvation from generation to generation."—ISAIAH LI.

Paley observes, that "the *metamorphosis* of insects from grubs into moths and flies, is an astonishing process. A hairy caterpillar is transformed into a butterfly. Observe the change. We have four beautiful wings where there were none before; a tubular proboscis, in the place of a mouth with jaws and teeth; six long legs, instead of fourteen feet. In another case, we see a white, smooth, soft worm, turned into a black, hard, crustaceous beetle, with gauze wings. These, as I said, are astonishing processes, and must require, as it should seem, a proportionably artificial apparatus.

Fig. 79.—THE PEACOCK BUTTERFLY.

The hypothesis which appears to me most probable, is that, in the grub, there exists at the same time three animals, one within another, all nourished by the same digestion, and by a communicating circulation; but in different stages of maturity. The latest discoveries made by naturalists, seem to favour this supposition. The insect, already equipped with wings, is descried under the membranes both of the worm and nymph. In some species, the proboscis, the antennæ, the limbs, and wings of the fly, have been observed to be folded up within the body of the caterpillar; and with such nicety as to occupy a small space only under the two first wings. This being so, the outermost animal, which, besides its own proper character, serves as an integument to the other two, being the farthest advanced, dies, as we suppose, and drops off first. The second, the pupa or chrysalis, then offers itself to observation. This also, in its turn, dies; its dead and brittle husk falls to pieces, and makes way for the appearance of the fly or moth. Now, if this be the case, or indeed whatever explication be adopted, we have a

"That which the palmer-worm hath left hath the locust eaten; and that which the locust hath left hath the canker-worm eaten; and that which the canker-worm hath left hath the caterpillar eaten."—JOEL I.

prospective contrivance of the most curious kind; we have organisations *three deep;* yet a vascular system, which supplies nutrition, growth, and life, to all of them together."

Lord Brougham, in a note upon the above, does not support Paley's view. He says "It is more than probable that the parts which are to appear in the perfect insect do *not* exist in the larvæ, where there is not much difference between the larva and pupa, excepting at the time just previous to its becoming a pupa, at which time the larva is motionless and torpid. The caterpillar of a moth, when about to turn into a pupa, provides for the protection of the latter state, either by surrounding itself with a web, or by some other means. Soon after this is accomplished, the caterpillar becomes motionless, or nearly so; it can neither eat nor crawl. At this time, and *not before*, the parts of the pupa are forming within the skin of the caterpillar, which may be easily seen by dissection."

It appears to the author, however, that Paley is partially right, and Lord Brougham totally wrong, in these remarks. When Lord Brougham asserts that the parts of the pupa are forming within the skin of the caterpillar at that time when the transformation begins, "and not before, which may be easily seen by dissection," he forgets, that although in some instances it is the first moment when, to the human eye, the organs of the new creature *become perceptible*, that the "*three deep*" nature which Paley attributes to the *grub*, must really have existed *in the egg*—that the *butterfly* originated *in the egg*, as certainly as did the *caterpillar*, or the *crysalis*, and that unless that egg had possessed its three mysterious embryos, it would have been impossible for the grub to have progressed to the stages of transformation. No one has ever known the embryo of a bird's egg to pass through three distinct and dissimilar states of existence; nor has any one ever known the embryo of the butterfly's egg to stop short at either of the stages, if the proper conditions of its existence and development were supplied to it. *Why?* Because the embryo of the insect has a *threefold* nature, while that of the bird is *single.*

"They shall cut down her forest, saith the Lord, though it cannot be searched; because they are more than the grasshoppers, and are innumerable."—
JEREMIAH XLVI.

CHAPTER LXVI.

1292. *Why does the caterpillar become torpid when passing into the state of the chrysalis?*

Because in all probability, where the difference between the first and the ultimate form is considerable, the organs of the insect having to undergo great changes, it would suffer considerable pain. Torpor comes upon the insect, it is thrown into a state similar to that of a person who has inhaled chloroform; and after what has, in all probability, proved a pleasant dream, the insect awakes to find itself changed and beautified.

1293. *Why are the pupæ of grasshoppers and other insects, when about to undergo transformation, still active and sensitive?*

Because, as there is but a *slight difference* between the form which they have in the pupa state, and that which they ultimately assume, they do not require the state of torpidity to save them from pain, nor to arrest their movements while their organs are being changed. With them *the outer skin is thrown off*, and they are then perfect insects.

1294. *Why do caterpillars, when about to pass through the chrysalis state, attach themselves to the leaves of plants, &c.?*

Because they know instinctively that for a time they will be *unable to controul their own movements, and to avoid danger.* They therefore choose secure and dry places, underneath leaves, or in the crevices of old and dry walls, and there they firmly attach themselves, to await the time of their liberation.

1295. *Why do insects attach their eggs, to leaves &c.?*

Because, as the eggs have to be preserved during the winter, the insect attaches them to some surface which will be a *protection to them.* Generally speaking, the eggs are attached to the permanent stems of plants, and not to those leafy portions which are liable to fall and decay. The spider *weaves a silken bag* in which it deposits its eggs, and then it hangs the bag in a sheltered situation. Nature

"Lay up for yourselves treasures in heaven, where neither moth nor rust doth corrupt, and where thieves do not break through and steal."—MATT. VI.

keeps her butterflies, moths, and caterpillars, locked up during the winter, in their egg-state; and we have to admire the various devices to which, if we may so speak, the same nature has resorted for the *security* of the egg. Many insects enclose their eggs in a silken web; others cover them with a coat of hair, torn from their own bodies; some glue them together; and others, like the moth of the silkworm, glue them to the leaves upon which they are deposited, that they may not be shaken off by the wind, or washed away by rain; some again make incisions into leaves, and hide an egg in each incision; whilst some envelope their eggs with a soft substance, which forms the first aliment of the young animal; and some again make a hole in the earth, and, having stored it with a quantity of proper food, deposit their eggs in it.

1296. *Why do butterflies fly by day?*

Because they are *organised to enjoy light and warmth*, and they live upon the sweets of flowers which by day are most accessible.

1297. *Why do moths fly by night?*

Because they are *organised to enjoy subdued light* and cool air; and as they take very little food during the short life they have in the winged state, they find sufficient by night. Some of the moths, like that of the silk-worm, take no food from the time they escape from the chrysalis until they die.

Because, also, they form the food of bats, owls, and other of the night-flying tribes.

1298. *Why are the bodies of moths generally covered with a very thick down?*

Because, as they fly by night, they are liable to the effects of cold and damp. The moths, therefore, are nearly all of them covered with a very thick down, quite distinguishable from the lighter down of butterflies.

1299. *Why do moths fly against the candle flame?*

Because their eyes are organised *to bear only a small amount of light.* When, therefore, they come within the light of a candle, their sight is overpowered and their vision confused; and as they cannot distinguish objects, they pursue the light itself, and fly against the flame.

1300. *Why do insects multiply so numerously?*

Because they form the food of larger animals, and especially of birds. A single pair of sparrows and a nest of young ones have been estimated to consume upwards of *three thousand* insects in a week.

1301. *Why does the "death-watch" make a ticking noise?*

Because the insect is one of the beetle tribe, having a horny case upon its head, *with which it taps upon any hard substance*, the ticking is the call of the insect to its species, just as the noise made by the cricket is a note of communication with other crickets.

There is a superstition connected with the death-watch, which, like most superstitions, is based upon the theory of *probabilities*. The death-watch is usually heard in the spring of the year, and a superstition runs to the effect that some one in the house will die before the year has ended. Persons who are superstitious are never very strict in the interpretation of their predictions; and therefore, whether a person dies in the house or out of it, in the same room where the death-watch was heard, or across the wide Atlantic, so that there be some kind of relationship, or even acquaintance, between the person who hears the omen, and the person dying, the event is sure to be connected with the prophetic sounds of the death-watch. Little weens the small timber-boring beetle, when he is tapping gently to call his mate, and perhaps peeping into every corner and crevice to find her, that he is sending dismay into the heart of some superstitious listener, who, in ignorance of a simple fact, overwhelms herself with an imaginary grief.

1302. *Why are insects in the first stage, after leaving the egg, said to be in the "larva" state?*

Because the term larva is derived from the Latin *larvated*, meaning masked, clothed as with a mask; the term is meant to express that the future insect is disguised in its first form.

1303. *Why are insects in the second state said to be in the "pupa" state?*

Because the term is derived from the Latin *pupa*, from a slight resemblance in the manner in which the insects are enclosed, to that in which it was the fashion of the ancients to *bandage their infants*.

1304. *Why are insects in the "pupa" stage also called "chrysalides?"*

Because, as the Latin term implies, it is adorned with gems. Many chrysalides are *studded with golden and pearl-like spots*.

"Thou hast set all the borders of the earth: thou hast made summer and winter."—PSALM LXXIV.

1305. *Why are the perfect insects said to be in the "nymph" state?*

Because their joyful existence, and their beautiful forms, give them a fancied resemblance to the *nymphs of the heathen mythology*. The nymphs were supposed goddesses of the mountains, forests, meadows, and waters.

This term has generally, but very improperly, been also applied to the pupa state, so that *pupa*, *chrysalis*, and *nymph* have all been employed to represent one state. This is obviously an error, as there is nothing in the condition of the *pupa* or *chrysalis* that can at all accord with the mythological idea of a *nymph*, and which, in reference to the beautiful and joyous fly, finds a much truer application.

CHAPTER LXVII.

1306. *Whence does the snail obtain its shell?*

Young snails come from the egg *with a shell upon their backs.*

1307. *How does the shell grow with the increase of size of the animal?*

The soft slime which is yielded by the body of the animal, *hardens upon the orifice of the shell*, and thus increases its size.

Fig. 80.—COMMON GARDEN SNAIL.

1308. *Why is the shell spiral?*

Partly because of its original formation; but also because, *as the shell grows*, the opening is elongated, and thrown up, causing the spiral body of the shell to turn, and so to wind its growth around the centre.

1309. *Why has the snail four tentacula attached to its head?*

Because the insect, having no other limbs, is provided with those projecting members, the lower two serving as *feelers* and the upper two also as *feelers* and *eyes*. These, projecting in the front of the animal, impart to it a consciousness of surrounding objects, and especially of those which lie in its path.

1310. *Why is the snail able to move, without feet?*

Because it has attached to its body a fringe of muscular skin, which is capable of considerable contraction and expansion, and by alternately stretching and shortening this, the snail is able to draw himself along.

1311. *Why do we see no snails in the winter time?*

Because they bury themselves in the ground, or in holes, where they remain *in a torpid state* for several months. Before they enter into the torpid state, they form with their slimy secretion, and with some earthy matters which they collect, a strong cement with which they seal up the opening to their shells.

1312. *Why can snails live in shells thus sealed?*

Because they leave, in the thin wall by which they close themselves in, *a small hole*, too small to admit water, but large enough to let in sufficient air to carry on their feeble respiration during their winter sleep.

1313. *Why do insects abound in putrid waters, and in decaying substances?*

Because they have been endowed with appetites and with constitutions that enable them to live upon and to enjoy corrupt matter. In this point of view the maggots of flies are exceedingly useful; a dead carcass is speedily threaded by them in every direction; thus that corrupt matter which, in a large mass, would poison the air, is taken up in small portions by millions of living bodies, and by them *dispersed*, and becomes innoxious.

"For he maketh small the drops of water: they pour down rain according to the vapour thereof."—JOB XXXV.

1314. *Why do we see, in tanks of rain water, insects rising to the surface?*

Because numerous insects pass through their first stages of existence in *water*, and among them the common gnat. The gnats of the previous season having deposited their eggs on the sides of the water-butt, the warm water developes them, and the larvæ of the gnats appear (Fig. 81; *c* natural size of larva; *b* larva magnified).

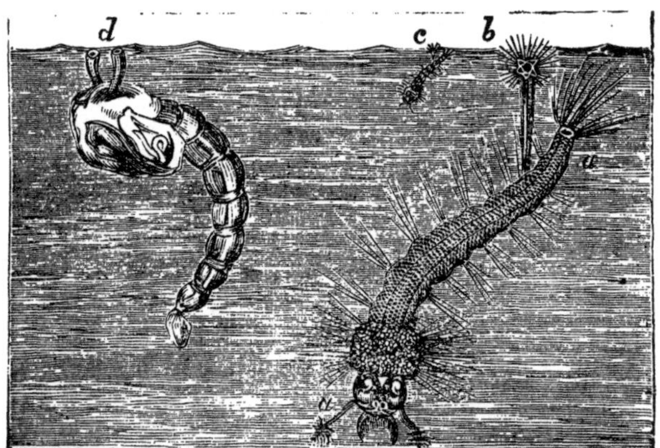

Fig. 81.—LARVA AND PUPA OF GNAT.
(*Greatly magnified.*)

1315. *Why do they continually rise to the surface of the water?*

Because they require to breathe air, and therefore they come up to the surface, where, elevating the tube *(b)* above the surface of the water, they are enabled to breathe.

1316. *Why do some appear to have larger heads than others?*

Those that have apparently larger heads, and that breathe through tubes attached to their heads *(d)* are in the ***pupa***, or second stage of development, and underneath the large shield by which their heads are marked, their wings, feet, &c., are being formed.

1317. *Why, when the water is disturbed, do the larvæ descend more rapidly than the pupæ?*

Because the pupæ are in a torpid condition, awaiting the formation of their perfect organs.

1318. *Why are the flies able to escape from the water?*

Because, as their formation becomes perfected, and the fluids of the body of the pupa become absorbed in the production of the light texture of the wings, &c., *the body and its case become lighter than the water*, and rise and float upon the surface. The pupa-case then forms a natural boat, from which the fly emerges, and spreading its wings, enters upon the final state of its existence

Fig. 82.—THE PERFECT GNAT, ESCAPING FROM THE PUPA-CASE.
(*Greatly magnified.*)

This interesting metamorphosis may be seen going on in the summer time, in every pond, brook, and reservoir. A fine sunny morning calls up millions of those little boats from beneath the surface, and the diver within that wonderful little bell breaks its sealed doors, and flies away to enjoy the bright sunshine.

1319. *Why are beetles denominated "coleoptera?"*

Because they have wings protected by horny sheaths; the term *coleoptera* signifies *wings in a sheath.*

" They shall lie down in the dust; and the worms shall cover them."—JOB XXI

1320. *Why have beetles hard horny wing-cases?*

Because they live underground, or in holes excavated in wood, &c. If, therefore, their wings were not protected by a hard and firm covering, they would be constantly *liable to destruction* from the movement of the insect within hard and rough bodies.

Fig. 83.—STAG-BEETLE, SHOWING ITS WINGS UNFOLDED, AND THE WING CASES OPEN.

The *elytra*, or scaly wings of the genus of scarabæus, or beetle, furnish an example of this kind. The true wing of the animal is a light, transparent membrane, finer than the finest gauze, and not unlike it. It is also, when expanded, in proportion to the size of the animal, very large. In order to protect this delicate structure, and, perhaps, also to preserve it in a due state of suppleness and humidity, a strong, hard case is given to it, in the shape of the horny wing which we call the elytron. When the animal is at rest, the gauze wings lie folded up under this impenetrable shield. When the beetle prepares for flying, he raises the integument, and spreads out his thin membrane to the air. And it cannot be observed without admiration, what a tissue of cordage, *i. e.* of muscular tendons, must run in various and complicated, but determinate directions, along this fine surface, in order to enable the animal, either to gather

"The Lord is good; his mercy is everlasting; and his truth endureth to all generations."—PSALM C.

it up into a certain precise form, whenever it desires to place its wings under the shelter which nature hath given to them, or to expand again their folds when wanted for action.

In some insects, the elytra cover the whole body; in others, half; in others only a small part of it; but in all, they completely hide and cover the true wings. Also,

Many, or most of the beetle species lodge in holes in the earth, environed by hard, rough substances, and have frequently to squeeze their way through narrow passages; in which situation, wings so tender, and so large, could scarcely have escaped injury, without both a firm covering to defend them, and the capacity of folding themselves up under its protection.

1321. *Why have many of the beetle tribe large strong horns?*

Because, as they live in holes in the earth, or in excavations in wood, they use their horns to *dig out their places of retreat.*

1322. *Why has the giraffe a small head?*

Because, being set upon the end of a very long neck, the animal would be *unable to raise it* if it were heavy.

1323. *Why has the giraffe a long neck?*

Because it *feeds upon the branches of tall trees.*

1324. *Why has the giraffe a long and flexible tongue?*

Because it is thereby enabled to lay hold of the tender twigs and branches, *and draw them into its mouth,* avoiding the coarser parts of the branches.

1325. *Why are the nostrils of the giraffe small and narrow, and studded with hairs?*

Because the hairs and the peculiar shape of the nasal passages are designed as a protection against the insects which inhabit the boughs of the trees upon which the giraffe feeds; and also against the sands of the desert, which storms raise into almost suffocating clouds.

1326. The distribution of animals, or *Zoological Geography,* is a of great interest, and should be carefully studied in connection with *Botanical Geography* (*see* 1208). The highest department of the animal kingdom (writes the Rev. W. Milner) commences with the class of *Birds,* which may be naturally divided into the three great orders of aerial, terrestrial, and aquatic. Aggre

15

gation into immense flocks is a distinguishing feature of several species, especially of the aquatic order, which form separate colonies, building their nests in the same state, though other spots equally adapted are at no great distance.

Fig. 84.—GIRAFFE FEEDING.

Hence the Vogel-bergs, or bird-rocks of the northern seas, one of which at Westmannsharn in the Faroe group of islands, seldom intruded upon by man, presents a most extraordinary spectacle to the visitor. The Vogel-berg lies in a frightful chasm in the precipitous shores of the island, which rise to the height of a thousand feet, only accessible from the sea by a narrow passage. Here congregate a host of birds. Thousands of guillemots and auks swim in groups

He rained flesh upon them as dust, and feathered fowls like as the sand of the sea."—PSALM LXXVIII.

around the boat which conveys man to their domain, look curiously at him, and vanish beneath the water to rise in his immediate neighbourhood. The black guillemot comes close to the very oars. The seal stretches his head above the waves, not comprehending what has disturbed the repose of his asylum, while the rapacious skua pursues the puffin and gull. High in the air the birds seem like bees clustering about the rocks, whilst lower they fly past so close that they might be knocked down with a stick. But not less strange is the domicile of this colony. On some low rocks scarcely projecting above the water sit the glossy cormorants, turning their long necks on every side, Next are the skua gulls, regarded with an anxious eye by the kittiwakes above. Nest follows nest in crowded rows along the whole breadth of the rock, and nothing is visible but the heads of the mothers and the white rocks between, A little higher on the narrow shelves sit the guillemots and auks, arranged as on parade, with their white breasts to the sea, and so close that a hailstone could not pass between them. The puffins take the highest station, and, though scarcely visible, betray themselves by their flying backwards and forwards. The noise of such a multitude of birds is confounding, and in vain a person asks a question of his nearest neighbour. The harsh tones of the kittiwakes are heard above the whole, the intervals being filled with the monotonous note of the auk, and the softer voice of the guillemot. When Graba, from whose travels this description is principally drawn, visited the Vogel-berg, he was tempted by the sight of a crested cormorant to fire a gun, but what became of it, he remarks, it was impossible to ascertain. The air was darkened by the birds roused from their repose. Thousands hastened out of the chasm with a frightful noise, and spread themselves over the ocean. The puffins came wandering from their holes, and regarded the universal confusion with comic gestures. The kittiwakes remained composedly in their nests, whilst the cormorants tumbled headlong into the sea. Similar great congregations of the feathered race appear where the shores are rocky high, and precipitous, but this is strikingly the case, where

—"The northern ocean, in vast whirls,
Boils round the naked melancholy isles
Of farthest Thule; and the Atlantic surge
Pours in among the stormy Hebrides.
Who can recount what transmigrations there
Are annual made? what nations come and go?
And how the living clouds on clouds arise?
Infinite wings! till all the plume-dark air
And rude resounding shore are one wild cry."

1327. Most terrestrial birds, unacquainted with man, exhibit a remarkable tameness, and are slow in acquiring a dread of him, even after repeated lessons that danger is to be apprehended from his neighbourhood. Mr. Darwin speaks of a gun as almost superfluous in the unfrequented districts of South America, for with its muzzle he pushed a hawk off the branch of a tree. Once, while lying down, a mocking thrush alighted on the edge of a pitcher, made of the shell of a tortoise, which he was holding in his hand, and began very leisurely to sip the water, even allowing him to handle it while seated on the vessel. In Charles Island, which had been colonised about six years, he saw a boy sitting by a well with a switch in his hand, with which he killed the doves and finches as they came to drink; and for some time had been constantly in the habit of waiting by the well for the same purpose, to provide himself with his dinners. In the Falkland Islands, at Bourbon, and at Tristan d'Acunha, the same tame-

"As a bird that wandereth from her nest; so is a man that wandereth from his place."—PSALM XXVII.

ness was noticed by the early visitors. On the other hand, the small birds in the arctic regions of America, which have never been persecuted, exhibit the anomalous fact of great wildness. From a review of various facts, Mr. Darwin concludes, "first, that the wildness of birds with regard to man is a particular instinct directed against him, and not dependent on any general degree of caution arising from other sources of danger; secondly, that it is not acquired by individual birds in a short time, even when much persecuted; but that in the course of successive generations it becomes hereditary. Comparatively few young birds in any one year have been injured by man in England, yet almost all, even nestlings, are afraid of him; many individuals, however, both at the Galapagos and at the Falklands, have been pursued and injured by man, but yet have not learned a salutary dread of him."

1328. Numerous species of birds may be regarded as the favourites of nature on account of the gracefulness given to their shape, and the richly-coloured plumage with which they are adorned, as evidenced in the gaudy liveries of many of the parrot tribe, and the forms and hues of the birds of paradise. But they are especially interesting to man for the faculty of song with which they are endowed; in some, "most musical, most melancholy," in others, sprightly and animating, inspiriting the sons of toil under the burdens peculiar to their station. It deserves to be remarked, as an instance of compensation and adjustment, that whilst the birds of the temperate zone are far inferior to those of tropical climes in point of beauty, they have far more melodious notes in connection with their less attractive appearance.

1329 From the powerful means of locomotion possessed by several of the bird tribe, and their great specific levity, air being admitted to the whole organisation as water to a sponge, it might be inferred, that the entire atmosphere was intended to be their domain, so that no species would be limited to a particular region. The common crow flies at the rate of twenty-five miles an hour; the rapidity of the eider-duck, *Anas mollissima*, is equal to ninety miles an hour; while the swifts and hawks travel at the astonishing speed of a hundred and fifty miles in the same time. It is true that some species have a very extensive range, as the nightingale, the common wild goose, and several of the vulture tribe. The same kind of osprey or fishing-eagle that wanders along the Scottish shores appears upon those of the south of Europe, and of New Holland. The lammergeyer haunts the heights of the Pyrenees, the mountains of Abyssinia, and the Mongolian steppes; and the penguin falcon occurs in Greenland, Europe, America, and Australia. In general, however, like plants and terrestrial quadrupeds, the birds are subject to geographical laws, definite limits circumscribing particular groups. The common grouse of our own country affords a striking exemplification of this arrangement, as it is nowhere met with out of Great Britain; and other examples occur of a very scanty area containing a species not to be found in any other region. The celebrated birds of paradise are exclusively confined to a small part of the torrid zone, embracing New Guinea and the contiguous islands; and the beautiful Lories are inhabitants of the same districts, being quite unknown to the New World. Parroquets are chiefly occupants of a zone extending a few degrees beyond each tropic, but the American group is quite distinct from the African, and neither of these have one in common with the parrots of India. The great eagle is limited to the highest summits of the Alps; and the condor, which soars above the peak of the loftiest of the Andes, never quits that chain. Humming-birds are

'There is a path which no fowl knoweth, and which the vulture's eye hath not seen."—JOB XVIII.

entirely limited to the western hemisphere, where a particular species is sometimes bounded by the range of an island, while others are more extensively spread, the *Trochilus flammifrons*, common to Lima, being observed by Captain King upon the coast of the Straits of Magellan, in the depth of winter, sucking the flowers of a large fuschia, then in bloom in the midst of a shower of snow. Among the birds incapable of flight, which rival the quadrupeds in their size, the intertropical countries of the globe have their distinct species, presenting similar general features of organisation, as the ostrich of Africa and Arabia, the cassowary of Java and Australia, and the touyou of Brazil. In the arctic regions, we meet with species peculiar to them, the *Strix laeponicus* or Lapland owl, and the eider-duck, an inhabitant of the shores, from whose nests the eider-down is obtained. Several families of maritime birds are likewise limited to particular oceanic localities. Approaching the fortieth parallel of latitude, the albatross is seen flitting along the surface of the waves, and soon afterwards the frigate and other tropical birds appear, which never wander far beyond the torrid zone. It thus appears, that, notwithstanding the great locomotive powers of birds, particular groups have had certain regions assigned to them as their sphere of existence, which they are adapted to occupy, and to which they adhere in the main, though it is easy to conceive of natural causes occasionally constraining to a migration into new and even distant territories. Captain Smyth informed Mr. Lyell, that when engaged in his survey of the Mediterranean, he encountered a gale in the Gulf of Lyons, at the distance of between twenty and thirty leagues from the coast of France, which bore along many land-birds of various species, some of which alighted on the ship, while others were thrown with violence against the sails. In this manner, many an islet in the deep, after ages of solitude and silence, uninterrupted except by the wave's wild dash, and the wind's fierce howl, may have received the song of birds, forced by the tempest from their home, and compelled to seek a new one under its direction.

1330. There is no feature more remarkable in the economy of birds than the periodical migrations, so systematically conducted, in which five-sixths of the whole feathered population engage. In the case of North America, according to an estimate by Dr. Richardson, the passenger-pigeons form themselves into vast flocks for the journey, one of which has been calculated to include 2,230,000,000 individuals. We are familiar with the cuckoo as our visitor in spring, and with the house-swallow as our guest through the summer, the latter usually departing in October to the warmer regions of the south, wintering in Africa, returning again when a more genial season revives its insect food. By cutting off two claws from the feet of a certain number of swallows, Dr. Jenner ascertained the fact of the same individuals re-appearing in their old haunts in the following year, and one was met with even after the lapse of seven years. The arctic birds migrate farther south, when the seas, lakes, and rivers become covered with unbroken sheets of ice; the swans, geese, ducks, divers, and coots flying off in regular phalanxes to regions where a less rigorous winter allows of access to the means of life. Hence, soon after, we lose the swallows, we gain the snipes and other waders, which have fled from the hard frozen north to our partially frozen morasses, where their ordinary nutriment may still be obtained. The equinoctial zone, where the seasonal change is that of humidity and drought furnishes an example of the same phenomenon. As soon as the Orinoco is swollen by the rains, overflows its banks, and inundates the country on either side, an innumerable quantity of aquatics leave its course for the West India islands on the north, and the valley of the Amazon on the south, the increased depth of the river, and the flooded state of the shores, depriving them of the usual supply of fish and insects. Upon the stream decreasing, and retiring within its bed, the birds return.

1331. A comparison between the quadrupeds of the Old and New Worlds is in every point strikingly in favour of the former. Not only has the western continent no animals of such giant bulk as those of the eastern, but no examples of such high organisation, such power and courage, as the African lion and the Asiatic tiger display. Buffon's remark must indeed be considerably modified, respecting the cowardice of the American feline race; for the jaguar of the woods about the Amazon, when attacked by man, will not hesitate to accept his challenge, will even become the assailant, nor shrink from an encounter against the greatest odds. The following passages from the writings of Humboldt show that this transatlantic animal is not to be despised:—

"The night was gloomy; the Devil's Wall and its denticulated rocks appeared from time to time at a distance, illuminated by the burning of the savannahs, or wrapped in ruddy smoke. At the spot where the bushes were the thickest, our horses were frightened by the yell of an animal that seemed to follow us closely. It was a large jaguar, that had roamed for three years among these mountains. He had constantly escaped the pursuit of the boldest hunters, and had carried off horses and mules from the midst of enclosures; but, having no want of food, had *not yet* attacked men. The negro who conducted us uttered wild cries. He thought he should frighten the jaguar; but these means were of course without effect. The jaguar, like the wolf of Europe, follows travellers even when he will not attack them; the wolf in the open fields and in unsheltered places, the jaguar skirting the road, and appearing only at intervals between the bushes."

The same illustrious observer also remarks,—

"Near the Joval, nature assumes an awful and savage aspect. We there saw the largest jaguar we had ever met with. The natives themselves were astonished at its prodigious length, which surpassed that of all the tigers of India I had seen in the collections of Europe."

Still these were extraordinary specimens of the race, and leave the fact undoubted, that the most formidable of the western *Feræ* has no pretensions to an equality with his congener, the tyrant of the jungles of Bengal.

1332. In vain also we look among the tribes of America for a rival in outward appearance to the giraffe, so remarkable for its height, its swan-like neck, gentle habits, and soft expressive eye; while of the animals most serviceable to mankind—the horse, the ox, the ass, the goat, and the hog—not a living example of either was known there before its occupancy by the Europeans. But, however inferior the animal race of the New may be as compared to those of the Old world, the balance between the two appears to have been pretty equal in remote ages; geological discovery has disproved the assertion of Buffon, that the creative force in America in relation to quadrupeds never possessed great vigour, and has established the fact, that it is only the more recent specimens of its energy that are upon an inferior scale. The relics of the unwieldly magatherium, of the gigantic sloth, and armadillo-like animals, discovered in great abundance imbedded in its soil, prove that at a former period it swarmed with monsters of equal bulk with those that now roam in the midst of Africa and Asia. The estuary deposit that forms the plains westward of Buenos Ayres, and covers the gigantic rocks of the Bando Oriental, appears to be the grave of extinct gigantic quadrupeds.

"But wild beasts of the desert shall lie there; and their houses shall be full of doleful creatures; and owls shall dwell there, and satyrs shall dance there."—ISAIAH XIV.

1333. There are various animals which are very widely dispersed, enduring the extremes of tropical heat and of polar cold, which are either in a wild condition or in a state of domestication. Wild races, considered to be varieties of the domestic dog, occur in India, Sumatra, Australia, Beloochistan, Natolia, Nubia, various parts of Africa, and both the Americas; while in subjection to man, the dog is his faithful companion, and has followed his steps into every diversity of climate and of situation to which he has wandered. The north temperate zone of the Old Continent appears to be the native region of the ox, which passes in Lapland within the arctic circle, and has been spread over South America since its first introduction by the Spaniards. The horse, originally an inhabitant of the temperate parts of the Old World, has shared in a similar dispersion, and now exists in the high latitude of Iceland, in the desolate regions of Patagonia, and roams wild in immense herds over the Llanos of the Orinoco, leading a painful and restless life in the burning climate of the tropics. Humboldt draws a striking picture of the sufferings of these gifts of the Old World to the New, returned to a savage state in their western location.

"In the rainy season, the horses that wander in the savannah, and have not time to reach the rising grounds of the Llanos, perish by hundreds amidst the overflowings of the rivers. The mares are seen, followed by their colts, swimming, during a part of the day, to feed upon grass, the tops of which alone wave above the waters. In this state they are pursued by the crocodiles; and it is by no means uncommon to find the prints of the teeth of these carnivorous reptiles on their thighs. Pressed alternately by excess of drought and of humidity, they sometimes seek a pool, in the midst of a bare and dusty soil, to quench their thirst; and at other times flee from water and the overflowing rivers, as menaced by an enemy that encounters them in every direction. Harassed during the day by gad-flies and mosquitoes, the horses, mules, and cows find themselves attacked at night by enormous bats, that fasten on their backs, and cause wounds which become dangerous, because they are filled with acaridæ and other hurtful insects. In the time of great drought, the mules gnaw even the thorny melocactus (melon-thistle), in order to drink its cooling juice, and draw it forth as from a vegetable fountain. During the great inundations, these same animals lead an amphibious life, surrounded by crocodiles, water-serpents, and manatees. Yet, such are the immutable laws of nature, their races are preserved in the struggle with the elements, and amid so many sufferings and dangers. When the waters retire, and the rivers return into their beds, the Savannah is spread over with a fine odoriferous grass; and the animals of old Europe and Upper Asia seem to enjoy, as in their native climates the renewed vegetation of spring."

1334. The first colonists of La Plata landed with seventy-two horses, in the year 1535, when, owing to a temporary desertion of the colony, the animals ran wild; and in 1580, only forty-five years afterwards, it had reached the Straits of Magellan. The ass has a more restricted range than the horse, not being capable of enduring so great a degree of cold, though usually far from being considered a delicate animal. To the warmer parts of the temperate zone, between the 20th and the 40th parallels of latitude, the ass seems best adapted, not propagating much beyond the 60th, and only occurring in a state of degeneration beyond the 52nd. The sheep and goat tribe are widely spread, equally supporting the extremes of temperature. According to Zimmerman, the *Argali* or *Moufflon*, the original race of sheep, still exists on all the great mountains of the two continents; and the *Capricorn* and *Ibex*, the ancestors of the common goat inhabit the high European elevations. From the 64th degree of north latitude the hog is met with all over the old continent, and also in the islands of the Indian Ocean, peopled by the Malay race; and since its introduction into the

New World, it has diffused itself over it, from the 50th parallel north as far as Patagonia. Originally the cat was not known in America, nor in any part of Oceanica; but it has now spread into almost every country of the globe. Among animals entirely wild, the most extensively diffused, are the fox, hare, squirrel, and ermine; but the species are different in every region of the world; nor is there perhaps one example to be found of a species perfectly identical naturally existing in distant localities of the earth.

Respecting the *internal constitution and heat of the earth*, differences of opinion, and some very wild speculation have existed. We find in Humboldt's "Cosmos" the following remarks:—

1335. "It has been computed at what depths liquid and even gaseous substances, from the pressure of their own superimposed strata, would attain a density exceeding that of platinum, or of iridium; and in order to bring the actual degree of ellipticity, which was known within very narrow limits, into harmony with the hypothesis of the infinite compressibility of matter, Leslie conceived the interior of the Earth to be a hollow sphere, filled with "an imponderable fluid of enormous expansive force." Such rash and arbitrary conjectures have given rise, in wholly unscientific circles, to still more fantastic notions. The hollow sphere has been peopled with plants and animals, on which two small subterranean revolving planets, Pluto and Proserpine, were supposed to shed a mild light. A constantly uniform temperature is supposed to prevail in these inner regions, and the air being rendered self-luminous by compression, might well render the planets of this lower world unnecessary. Near the north pole, in 82 deg. of latitude, an enormous opening is imagined, from which the polar light visible in Aurora streams forth, and by which a descent into the hollow sphere may be made. Sir Humphry Davy and myself were repeatedly and publicly invited by Captain Symmes to undertake this subterranean expedition; so powerful is the morbid inclination of men to fill unseen spaces with shapes of wonder, regardless of the counter-evidence of well-established facts, or universally recognised natural laws. Even the celebrated Halley, at the end of the 17th century, hollowed out the earth in his magnetic speculations; a freely rotating subterranean nucleus was supposed to occasion, by its varying positions, the diurnal and annual changes of the magnetic declination. It has been attempted in our own day, in tedious earnest, to invest with a scientific garb that which, in the pages of the ingenious Holberg, was an amusing fiction."

The following are among the speculations which Humboldt thus severely but justly condemns:—

"The increase of temperature observed is about 1 deg. Fahr. for every fifteen yards of descent. In all probability, however, the increase will be found to be in a geometrical progression as investigation is extended; in which case the present crust will be found to be much thinner than we have calculated it to be. And should this be found to be correct, the igneous theory will become a subject of much more importance, in a geological point of view, than we are at present disposed to consider it. Taking, then, as correct, the present observed rate of increase, the temperature would be as follows:

Water will boil at the depth of 2,430 yards.
Lead melts at the depth of 8,400 yards.
There is red heat at the depth of 7 miles.
Gold melts at 21 miles.
Cast iron at 74 miles.
Soft iron at 97 miles.

And at the depth of 100 miles there is a temperature equal to the greatest artificial heat yet observed; a temperature capable of fusing platina, porcelain, and indeed every refractory substance we are acquainted with. These temperatures are calculated from Guyton Morveau's corrected scale of Wedgwood's pyrometer; and if we adopt them, we find that the earth is fluid at the depth

"He hath filled the hungry with good things; and the rich he hath sent empty away."—LUKE I.

of 100 miles from the surface, and that even in its present state very little more than the soil on which we tread is fit for the habitation of organised beings."

The above is to be found in Mr. Timbs's "Things not Generally Known," a little book which professes to set people right upon points on which they are in error!

Upon this subject Mr. Hunt, in his "Poetry of Science," says:—

1336. "A question of great interest, in a scientific point of view, is the temperature of the centre of the earth. We are, of course, without the means of solving this problem; but we advance a little way onwards in the inquiry by a careful examination of subterranean temperature at such depths as the enterprise of man enables us to reach. These researches show us, that where the mean temperature of the climate is 50 deg., the temperature of the rock at 59 fathoms from the surface is 60 deg.; at 132 fathoms it is 70 deg; at 239 fathoms it is 80 deg.; being an increase of 10 deg. at 59 fathoms deep, or 1 deg. in 35.4 feet; of 10 deg. more at 73 fathoms deeper, or 1 deg. in 43.8 feet; and of 10 deg. more at 114 fathoms still deeper, or 1 deg. in 64.2 feet.

Although this would indicate an increase to a certain depth of about one degree in every fifty feet, yet it would appear that the rate of increase diminishes with the depth. It appears therefore probable, that the heat of the earth, so far as man can examine it, is due to the absorption of the solar rays by the surface. The evidences of intense igneous action at a great depth cannot be denied, but the doctrine of a cooling mass, and of the existence of an incandescent mass, at the earth's centre, remains but one of those guesses which active minds delight in."

Upon the subject of *hunger* and *thirst*, by which living creatures are prompted to feast upon the bounties of nature, Sir Charles Bell says, in "Appendix to Paley's Natural Theology:"—

1337. "Hunger is defined to be a peculiar sensation experienced in the stomach from a deficiency of food. Such a definition does not greatly differ from the notions of those who referred the sense of hunger to the mechanical action of the surfaces of the stomach upon each other, or to a threatening of chemical action of the gastric juice on the stomach itself. But an empty stomach does not cause hunger. On the contrary, the time when the meal has passed the stomach is the best suited for exercise, and when there is the greatest alacrity of spirits The beast of prey feeds at long intervals; the snake and other cold-blooded animals take food after intervals of days or weeks. A horse, on the contrary, is always feeding. His stomach, at most, contains about four gallons, yet throw before him a truss of tares or lucerne, and he will eat continually. The emptying of the stomach cannot, therefore, be the cause of hunger.

"The natural appetite is a sensation related to the general condition of the system, and not simply referable to the state of the stomach; neither to its action, nor its emptiness, nor the acidity of its contents; nor in a starved creature will a full stomach satisfy the desire of food. Under the same impulse which makes us swallow, the ruminating animal draws the morsel from its own stomach.

1338. "Hunger is well illustrated by thirst. Suppose we take the definition of thirst—that it is a sense of dryness and constriction in the back part of the mouth and fauces; the moistening of these parts will not allay thirst after much fatigue or during fever. In making a long speech, if a man's mouth is parched, and the dryness is merely from speaking, it will be relieved by moistening, but if it comes from the feverish anxiety and excitement attending a public exhibition, his thirst will not be so removed. The question, as it regards thirst, was brought to a demonstration by the following circumstance. A man having a wound low down in his throat, was tortured with thirst: but no quantity of fluid passing through his mouth and gullet, and escaping by the wound, was found in any degree to quench his thirst.

"Thirst, then, like hunger, has relation to the general condition of the animal system—to the necessity for fluid in the circulation. For this reason, a man dying

Let us hear the conclusion of the whole matter; Fear God, and keep his commandments: for this is the whole duty of man."—ECCLESIASTES XII.

from loss of blood suffers under intolerable thirst. In both thirst and hunger, the supply is obtained through the gratification of an appetite; and as to these appetites, it will be acknowledged that the pleasures resulting from them far exceed the pains. They gently solicit for the wants of the body; they are the perpetual motive and spring to action."

Our task draws near to a conclusion; and we hope that those who have followed our teachings will thirst after further knowledge; that they will henceforward regard the great Book of Nature as the work of an Almighty Hand, and endeavour to find, for everything that Nature does, the *Reason Why*.

A high perception of the wisdom of the Divine Being, must necessarily be the result of an intelligent contemplation of the Divine works. To the ignorant, the name of God is an unmeaning word; it may inspire fear, but it does not develope love. To the dark mind of the untaught man, God is no more than one of those mysterious existences that awe the superstitious, and deter the wicked. There is no grafting of the soul of the man upon the eternal love. But knowledge brings man into communion with that Almighty wisdom which is the fountain of all truth and happiness. To the enlightened man, God is the sun of all goodness, around whom the attributes of Power, Wisdom, and Love, radiate and fill the universe. As man's physical eye cannot withstand the light of the sun, neither can man's spiritual eye see the whole glory of God. But as we can rejoice in the sunshine, and interpret the mission of the sunbeam, so can we find happiness in the Divine presence, and gather wisdom by the contemplation of the Creator's works.

Nature is a great teacher. What a lesson may be gathered from the germination of a seed; how uniformly the germs obey their destiny. However carelessly a seed may be set in the ground, the germ which forms the root, and that which is the architect of the stem, will seek their way—the one to light, the other to darkness—to fulfil their duty. The obstruction of granite rocks, cannot force the rootlet upward, nor drive the leaflet down. They may kill the germs by exhausting their vital powers in an endeavour to find the proper elements; but no obstruction can make a single blade of grass do aught but strive to fulfil the end for which it was created. Would that man were equally true to the purpose of his existence, and suffered neither the rocks of selfishness, nor the false light of temptation, to force or allure him from duty to his God.

THE END.

www.ingramcontent.com/pod-product-compliance
Lightning Source LLC
LaVergne TN
LVHW010208110826
845151LV00002B/638

9781425535728